Dangerous Waters

Tales of the Sea

By J. Gordon Mumford

Zebra Publishing House
New Westminster BC

zebra PUBLISHING HOUSE

Zebra Publishing House
413 - 3 Rialto Court
New Westminster, BC
Canada - V3M 6P2
Tel: 604-520-1487; Fax: 604-520-1497
E-mail zebrapage@shaw.ca

ISBN 978-0-9736297-2-9

Printed and bound in Canada. Layout by Zebra Page Design.

Cover photo: Convoy of merchant ships assembling in the North Atlantic. (National Archives of Canada, PA-115005).

Canadian Cataloguing in Publication Data

Mumford, J. Gordon, 1925-

Dangerous waters: tales of the sea / by J. Gordon Mumford.

Includes index.

ISBN 978-0-9736297-2-9

1. Mumford, J. Gordon, 1925-
2. World War, 1939-1945--Personal narratives, British
3. Seafaring life.
4. Merchant Mariners--Great Britain--Biography
5. Radio operators--Great Britain--Biography
6. World War, 1939-1945--Personal narratives, Canadian

I. Title

D811.M85 2008 **940.54'8141** **C2008-901003-5**

Contents

J. Gordon Mumford, Radio Officer

The *Soborg* and Iceland

No cooling breeze flows from the whirring fan in the radio cabin. The rotating curved blades bite into the humid air, circulating the cloying smell of vomit intermingled with the rancid odour emanating from the steam radiator.

The ship pitches violently, rising and falling in heavy seas. Securely fastened to the deck by a chain, my chair rises and falls with the ship. Waves of nausea surge through my body, and I reach for the red can.

The can is quite ordinary, ten inches tall and about six inches in diameter. When I first saw it in the kitchen, it contained cooking fat, but I persuaded the cook to give it to me.

"Sure you not want a bucket, Sparks?" he had said, in his strange broken English.

"No, this can has a lid on it," I replied. "That will keep the smell in. Anyway, it'd be just my luck for a bucket to tip over in this bloody weather."

The memory dissipates as another spasm of agonizing pain twists my body. When I throw up, however, I do gain some momentary relief. With the lid clamped firmly on the can, I am no longer aware of the odious smell.

The incessant chatter of Morse code flows from the headphones clamped to my head, drilling into my brain. Shakily clutching my pencil, I resume writing, automatically recording the stream of characters coming from the marine radio transmitter at Rugby. The messages are sent in cypher groups of five. Until decoded, they are gibberish, a meaningless jumble of letters and numbers.

The four hours of my watch pass slowly, and my watch seems endless. Another quick glance at the clock on the bulkhead reveals that finally it's almost twelve hundred hours. The Chief Sparks should be relieving me soon. Stretching my arms, I lay down the pencil. My limbs feel weak and shaky, but the thought of food renews my nausea, and I decide to skip lunch.

The heavy steel outer door opens with a clang, and a blast of cold air sends a shudder through my body. "Christ," the Chief says, grimacing as he enters, "the place stinks. Phew, what with the bloody radiator and that can of yours, we're in for a bad trip." Bluff in both speech and manner, the Chief Radio Officer is a north countryman of medium height in his late thirties.

Wearily I shake my head. "Chief, if I'd known life at sea was like this, I'd have stayed home."

His face breaks into a grin. "You'll get used to it. It'll take a few days, but then you will be all right. If you think this is bad, just wait until we get out into the Atlantic."

Outside, the wind is icy against my face, but it clears my head. Going to the leeward side of the ship, I empty the red can over the side. Tired out, I just want to get into my bunk, but my sleeping quarters are partitioned off from the radio cabin. My bunk can only be accessed by going through the radio cabin. I live and sleep within a few paces of my work.

By the time I return, the Chief has aired the radio cabin, and it is now chilly but inhabitable. The earphones are glued to his ears, and he is concentrating on incoming traffic. He doesn't even look up as I stumble past his back and enter my sleeping quarters.

Wedged into my bunk, I let my eyes rove over my new home. The once cream bulkheads are a dirty grey through years of exposure to coal dust that permeates the air and eats away at the paintwork. The *Soborg* is a collier, en route to the Tyne River for a cargo of coal needed to fuel the coal-burning boilers. Except for the fact that the ship is Danish, her resemblance to the ship in the Masefield's poem is remarkable.

> Dirty British coaster with a salt-caked smoke stack
> Butting through the Channel in the mad March days,
> With a cargo of Tyne coal...

The *Soborg* is a prewar relic, a coal-carrying cargo ship with a sturdy riveted hull but poor accommodation. My bunk is a narrow shelf, with two drawers underneath. A narrow cupboard is bolted to the bulkhead next to a small washstand that has a mirror above. A steam radiator, rumbling and gurgling like a tormented stomach, completes the picture.

The horizontal position calms my queasy stomach, still aching and sore from retching. I wonder if I will ever get used to the pitching and rolling of the ship. *Why did I want to go to sea?* Gloomily, I think of the months –no, years—that lie ahead. Depression sets in, and I stir restlessly in my bunk.

In another four hours, I will be on watch again. *This is my first ship and it has to be a two-watch system!* Only the Chief Radio Officer and myself share the radio watches. *Not only must I suffer from seasickness, but also from a twelve-hour work day, with four hours off and four on.*

The *Soborg* left the Green Hythe buoys in the Thames at first light this morning--September 24, 1942--heading north, and hugging the eastern seaboard of the English coast. We're bound for North Shields, a small port on the mouth of the river Tyne, and should be there tomorrow.

Lying in my bunk, my face burns with shame when I recall the incident with the starboard light. This morning, my face had been a ghastly white when I dashed for the side of the bridge. Convulsed with violent retching, I had drooped over the parapet, not caring whether it was port or starboard, windward or leeward. Unfortunately I made the wrong choice.

When my nausea subsided, I sighed with relief, but then I winced when I saw what I had done. The light was no longer green. "Oh, no," I groaned, "I've been sick all over the starboard light." Hastily, I moved away, hoping that no one had seen me.

The wheelhouse door had opened against the blustering wind and slammed shut as the Captain, his binoculars in hand, came out onto the bridge. He moved over to the wing and stood still for a moment, his body leaning against the bulwark as he read the signal flags of the commodore ship, which was pitching and rolling in heavy seas. Then, for some reason he glanced down and saw the starboard light. His startled face changed expression. Flushed with anger, he roared, "Who's been sick over the bloody starboard light."

Propped up against the bridge bulwark, I cringed when I heard his voice. It pierced my sickened senses as I stared blearily at the heaving and tossing North Sea.

"Sparks," he yelled. "Did you do that to my god-damned light?" He could only mean me. *Sparks* is the name coined aboard ship for radio men, like the term *Chippy* for a carpenter.

Nodding weakly, I muttered, "Sorry, Sir, just couldn't help myself. I, I feel awful."

He glowered at my cowering figure, with not an ounce of pity for the wretched youth in front of him. "Get a bucket of water and a mop, and clean that bloody light," he grunted, "and next time if you've got to be sick, use the leeward side of the vessel."

Groaning to myself, I hurried to fetch a mop and bucket. Then, dangling precariously over the side of the bridge, I struggled to clean the offending light. This was my first sea voyage and my first day at sea, but it was a bad start to the budding sea career of a young radio officer.

My seasickness is abating as I lie on my bunk, and my mind goes back over events of the past few days. *Is it really only two or three days ago since I had joined the ship? It seems longer.* I realize now that I should have known from the wording of the company's telegram that the ship was carrying coal. In fact, it said that the ship was a collier, but I was a country boy and the word meant little to me. If I had known how dirty colliers were, I would not have brought my new uniforms aboard.

When I arrived at the dock yard gate, the policeman at the entrance had examined my papers and passed me through. "She's lying alongside the docks, mate," he had said, a good natured grin on his face. "You can't miss her. She's in A12 dock, just past the tanker, about three or four hundred yards from here." There was a twinkle in his eye as he added, "She's dirty as hell."

My heart sank when I saw the ship. She was old and dirty. The grimy battleship grey paint peeled in places to reveal the red lead undercoat, giving her a mournful air. Her cargo holds, for'ard and aft of the bridge and mid section accommodation, were covered and sealed with their hatch covers and tarpaulin overlays clamped in place. She looked ready to go to sea.

From my bunk, I can hear the Morse traffic coming over the radio equipment in the next room where the Chief is on watch, and probably recording the shipping broadcasts from Rugby.

When I remember my first meeting with the Chief, I realize that I must be awfully green. *What was it he said?* "Didn't the office tell you that this was a collier, son? You'll ruin your uniform with all the coal dust around. It gets into everything. Look, you don't live too far away, do you? If I were you, I'd go home and change into some old clothes."

Sadly, I agreed. I was proud of my new uniform, but it was obviously unsuited to this old tub.

My thoughts wander on. Mother was surprised to see me back so soon. I chuckle as I remember the surprise on her face when I came in the door.

We talked as I repacked my suitcase. "It's a collier, Mum," I said. "There's coal dust everywhere. The Chief tells me it gets into everything. I'll need my old sweaters, socks and those short-sleeved shirts. Yes, the balaclava, too."

The ship lurches and I pull my knees up against the bunk

board. For a moment, I think about my father. The memories are fresh and painful because it is barely four weeks ago that he died. *So much has happened since then.*

While I was home, I had changed into a navy blue battle dress because it is more practical for a collier. I leave my bridge coat behind, and take a navy blue raincoat in its place. I had noticed that there were duffle coats on board in case we need them. I didn't get back to the ship until late that night.

I hear the Chief coughing. The deep rasping sound breaks my train of thought and brings me back to the present. I scarcely notice the constant dull thump-thump from the ship's engines in the background, but I doubt whether I will ever get accustomed to the pitch and roll of the ship. The steam radiator has a stale, sickening odour that assails my nostrils, but the sick feeling has gone. It's restful lying flat in my bunk, but I know the queasiness is there, waiting, just waiting for me to get up.

Trying to sleep, I focus on a black streak above my head, trying to exclude all other thoughts, but it's no use. Random scenes flicker through my head as my thoughts jump indiscriminately from one thing to another. I see the Captain and company officials sitting opposite me at a table in the Captain's cabin, as I sign the ship's articles. My mind moves on. The ship moves from the dock to the Green Hythe buoys in the river. We waited there to catch the early morning tide. That was only last night, but it seems an eternity since then.

A few hours ago, amidst incoming reports of enemy E-boat and aircraft activity, the *Soborg* sailed into the choppy swells of the Thames estuary, dropped the pilot, and moved into convoy formation. Within the short space of half an hour, I discovered the agonies of seasickness, the plague of those unfortunates who sail the sea.

Well, I think to myself as I draw the warm blankets closer round my body, *even Lord Nelson used to get seasick, and look how famous he became.* Drowsily, I wonder how he coped. *Perhaps he gave his orders lying down on a couch.* This idea gives me some comfort, and I close my eyes and go to sleep.

* * *

The shrill clangor of the ship's klaxon wakes me in the late afternoon. The noise shatters the patterned thud of the engines and the creaks and groans that the vessel makes as she pitches and rolls in the wintry seas. The strident tones echo along passageways, bringing the ship to maximum alert. Grabbing a tin hat, I leave the cabin, and scramble up an iron ladder to the armoured machine gun nest above the radio room. This is really a day of firsts: this will be my first day on a gun post.

The Chief is still on watch and, in the event of an attack, it is my responsibility to man the midships machine gun. We are halfway up the coast between the Thames estuary and the north country port of North Shields. An attack by enemy aircraft and possibly E-boats is expected. This area is known as E-boat alley because those high speed motor torpedo boats operate out of the occupied channel ports. I can feel the tension in the air as we wait for an attack to develop.

I throw off the waterproof canvas cover revealing an ancient Thompson machine gun on a swivel mounting, and check the round ammunition pan feeding the gun. Having been brought up in the country, I am used to guns, and did well on the three-week gunnery course prior to joining the *Soborg*. Perhaps today I can put the training given by the Royal Navy to good use. The gun feels good in my hands, and I feel that, if we are attacked, I can at least fight back.

Scanning the surrounding sea and overcast sky, I peer through the gun sights, focusing on imaginary targets. In some ways, war is still a game to me. I feel at one with the elements. The raw cold wind chills my face and body, and my seasickness has abated. *This is better*, I think, *more like the action I've waited for.*

Tensions are running high among the crew, and for good reason. Two months ago, the *Soborg* had been bombed and badly damaged in these waters. Hans, an elderly engineer, had told me about it last night in the wardroom.

Hans always calls me *Jong,* which means a young person. I am just seventeen and the youngest one on the ship. Perhaps he calls me that because I remind him of his sons, if he has any, or younger relatives, or maybe it is simply a paternal instinct.

"*Ja, loister, Jong,*" Hans said, idly watching his hands lying on the wardroom table. "Dey bombed us in de Thames Estuary. De bloddy bastard hit us in de stern, right op de tventy fife pounder gon. Many men killed. Ja, alles de gon crew and some of my stokers. Dey vere killed below decks ven de ammunition she blow up. Dey could not get out of de port holes because dey vere too small. De men died screaming. Dit vas awful. Dey died half in and half out of de port holes, poor bastards."

He paused a moment and then turned to me. "Vell, Jong, now ve ave got big port holes. Too damn late vir de odders, but maybe ve is ok dis time. Vat is dit dat jo Engels say? Lightning she no strike same place twice." He laughed at his own sardonic humour. I hope that he is right.

The ship had been in dry dock for repairs before I joined her. The port holes are now large enough for a man to scramble through to the well deck.

False alarm! The enemy aircraft must have been looking for other targets and had not spotted us in the overcast weather conditions. The crew breathes easier and the convoy slips away in the gathering dusk, hidden from searching eyes by rain squalls.

Early the next morning, we enter the Tyne and the short voyage to North Shields is over. This is now the home port of the *Soborg*, down river from the bustling City of Newcastle. Within two or three days, we are ordered to the nearby port of Blythe, a small seaport a few miles north of Newcastle, where we take on a cargo of coal.

As we load at the wharf, the air is thick with coal dust, and rumours about our final destination are also spreading rapidly through the ship. The Russian convoys to Murmansk are on everyone's mind. The Germans hold Norway, and Narvik is close to the northern route with airfields and enemy naval forces. *No, nobody likes to think of the Russian run as a destination.* Survival in the frigid waters of the Arctic is just a few minutes before hypothermia and death set in. *Christ*, I thought, *that would be murder*.

Fully loaded with coal, the *Soborg* moves out beyond the harbour breakwater into the anchorage, which is protected against enemy attacks by minefields. The ship swings idly on its anchor with the changing tide as we wait for orders to sail. Summoned by Naval Command to a convoy conference ashore, the Captain sets off in a naval launch soon after we anchor, and will return later with our orders.

A naval launch returns the Captain to the ship in the late afternoon, and then veers off to deliver other Captains to their ships in the anchorage. Unsmiling, the Captain hurries up the gangway and disappears into his quarters aft of the bridge. For security reasons, orders and convoy documentation are always opened in an anchorage or after the ship has put to sea. The Mate is summoned to the Captain's stateroom presumably to discuss the orders with the Captain.

A strange remote figure, middle aged with thinning hair, the Captain of the *Soborg* keeps to himself. His meals are served to him in his stateroom and he dines alone, not with his fellow officers. I seldom see him, except on the bridge and in the chartroom, or whenever he goes ashore for naval conferences or other business meetings. I never really get to know him, but, after being sick over his starboard light, I'm sure that he knows of me.

During supper that evening, the Mate breaks the news of our destination, although somehow it has already leaked out before that. "We are going to Iceland," the Mate says, "and not to Murmansk. The coal we have loaded is for refuelling a convoy bound for Russia. We are to rendezvous with the convoy in a fjord near the capital, Reykjavik."

Overcast skies and dirty weather conditions that have dogged us along the coast from the south continue. The weather is a mixed blessing. It helps to hide us from the enemy, but makes the task of keeping the convoy together difficult. Convoys sail at the speed of the slowest vessels, and if a ship can't keep up, she is left behind undefended to fend for herself.

My seasickness persists, despite predictions that in a few days I will get used to it. True, at times it does seem to ease a little. If I try not to eat, I still vomit, but my stomach is empty, and I simply bring up bile.

The Chief Sparks and I are the only Britishers aboard and the other officers and crew are all Danish or Dutch. My only contact with them is either on the bridge or in the officers' wardroom, a small room that also serves as the mess. Because I have no smattering at all of Danish, the conversation relies on their ability to speak English. A favourite topic at the meal tables is my seasickness. Their advice is endless.

"You know, Gordon," one says, "you must eat dry toast; that very good for stomach."

Another officer glances up, his mouth full of food, which he is munching with evident delight. "Ja, dat's true, *jong*, and jo must tighten jo belt. Dat vay jo vill not feel de honger so much."

On one occasion, an engineer whose English is far superior to the others, comments, "Yes, Gordon but you must also eat something. Try this curry; it's great. We Danes always eat plenty of curry."

I look at the horrible mess in the plate. The smell alone makes me nauseous. My stomach surges within me and, looking green, I bolt outside followed by amused laughter from the mess room.

Trouble seems to be my middle name on that ship, but I

am gaining experience and hopefully turning from a raw landlubber into a seafarer. I still have a long way to go. Visual signalling communications, such as semaphore, flags and Aldis lamp signalling, had not been a part of my training at the radio college. This deficiency becomes embarrassingly apparent only a few hours after leaving Blythe en route to Iceland. A naval escort vessel suddenly bears down on us out of the overcast, the seas tossing over its bow as it approaches. From its bridge an Aldis lamp flashes our name in Morse.

"Sparks," the Mate yells to me from the bridge as I stand on the deck. "Come up on the bridge and take this message."

I scramble up the steel gangway to the bridge and gingerly hold the lamp. I look at the Mate. "I've never used a lamp before Mr. Mate, but I'll try."

Aiming the lamp at the escort vessel and with my finger on the signalling trigger, I send the "K" signal, a dash followed by a dot and a dash, telling him to go ahead with his message. The naval lamp flashes back in a stream of long and short flashes, but not being used to visual signalling, I can't make any sense of the message. In desperation I send the wait signal, "AS" in Morse, and admit defeat.

"I'm sorry, Mr. Mate, but I can't make sense out of the message. Would you like me to call the Chief Sparks?"

He agrees and I hurry to the radio cabin, and get the Chief to help out his raw underling. My pride is hurt. I've lost face again and, together with having to cope with seasickness, life begins to lose its sheen. I feel overawed by these older, experienced men. My confidence is damaged, and I badly need an encouraging word to revive my depressed spirits.

I learn, of course; I have to. Sheltered in the lee of the midships accommodation, I watch the long and short flashes of the signalling lamps on ships communicating with each other. Gaining confidence with practice, I begin to read their messages. The exercise is interesting and I learn quickly. The Morse is the same, only now it is visual, not aural. The messages are routine, passing little bits of news and information between the ships. They are mainly on course changes, orders to reduce smoke, and other matters of interest to the convoy.

Learning to read these messages has another effect. The crew members see me reading the messages and speak to me. "What's happening, Sparks? What are they saying?" It salvages my pride to be able to translate the language of the talking lamps for them. I feel useful at last, and that is something I need. These visual communications links are our main means of communicating within the convoy because radio transmissions are forbidden unless we are under attack.

Rounding the top of Scotland between Inverness and the Orkney Islands, we run into gale force winds and seas, but forge slowly ahead on a bearing north northwest to Iceland. I suppose everybody has one of those most embarrassing moments of their lives, but I seem to excel in them. One such incident happens during the gale that is still raging as we alter course and leave the rugged coast of Scotland behind.

The seas seem higher than the mast head in the gale-lashed seas, although this is probably an optical illusion. The ship is rolling and pitching violently. Waves tower above the vessel; then, as she slips sideways, they crash down over the cargo deck. I cling to the aft deck railings in the lee of the midships accommodation, and watch the waves smash across the battened-down cargo hatches. The ship shudders and lurches, before returning tons of sea water in cascading masses, over and through openings in the cargo deck bulwarks.

My mind is not completely on the sea, however, but on the door that I am watching. That door opens into the ship's head, or, in everyday language, the toilet. The ship's head on this old collier doesn't have an inner door. Instead, the door opens directly out onto the railed walkway. The ancient plumbing has broken down, however, and I can't get the toilet to flush. In our home we have an outside toilet, consisting of a large wooden box with a hole cut in the top. Below is a pail that my father emptied once a week. The ship's toilet is new to me.

I vacillate outside the door, unable to make up my mind whether to go and find a pail or to stay and guard the door. Time decides the matter for me in the worst possible way. It becomes a comedy of errors.

First my Chief comes round the walkway, and I hurry to talk to him. "Chief. I've just been into the toilet, and...." I break off hurriedly at that point and, with dismay blanching my face, I see the Third Engineer disappear into the toilet before I can stop him.

He comes out in a hurry, and glares. There is anger in his voice. "Who the hell didn't flush the bloody toilet."

I freeze in dismay, and open my mouth to explain, but the words will not come out.

The Chief looks at me. "Gordon, you just told me that you were in the toilet. Was it you?"

My heart is in my boots and I wish the deck would swallow me up. "Ye-ye-yes," I stutter, "bu-bu-but I was trying to stop anyone from going in. The Third went in before I could stop him. There was no bucket, and I didn't like to leave the door unguarded." They simmer down, perhaps understanding, but it is a bad day for me. I hurry off to find a pail and rope in the galley.

Will I never get it right? I ask myself. I have to grow up fast and learn as I go along. Yes, I learn a lot on the *Soborg*; I will not forget those experiences in a hurry.

The temperature drops dramatically as we plough ever northward. The ice builds up thick on our superstructure and icicles hang down from surfaces where partial melting has taken place. Ice also forms on the steel mast stays, doubling or tripling their thickness. I am fascinated by the scene and stand on the deck outside the main midships accommodation. An engineer is busy with the deck crew trying to chip the ice off the outside water pumps to get them operational. As he chips the thick ice away from the pump handle and mechanism, he glances up and sees me watching with evident interest.

"Christ, Sparks," he says, "it's bloody cold here. Just look at all this ice."

I have to agree with that, because it is bitterly cold. To keep warm, we wear our warmest clothing underneath duffle coats and mittens provided by the ship. The duffle coats, with their protective hoods, are an absolute necessity when working outside in these temperatures.

Nearing the Arctic Circle, the snow-clad shores of Iceland rear up on the horizon. As the convoy approaches the rugged mountainous island, flags ordering an alteration in course flutter from the mast of the commodore's ship. The convoy is now running parallel with the coast as we head for Reykjavik, the capital and port. When the convoy anchors in the bay, the brightly coloured roofs stand out against the snow-covered ground and mountains.

The Mate is with the deck crew on the fo'c'sle head. On a shouted command from the bridge, the anchor holding pin is struck. The heavy chain links pass through the anchor port as the anchor plunges with a roar into the sea. We have arrived. Held fast by the anchor, we wait for further orders.

Iceland is the first foreign land that I have seen. I stare at it from the bridge with binoculars, intent on drinking in every detail with my eyes. I feel awe and excitement flood through my body and I long to go ashore and mingle with the people of this strange land. My dreams of travel and adventure are being answered; here is a new world.

As we lie at anchor, the crew throw long lines over the side with fishing hooks attached every two feet or so. I am amazed at the number of fish that we catch in the bay. Every line is alive with fish dangling on the hooks almost as soon as the lines hit the water.

Curious about how the galley will handle all these fish, I make my way to the lower deck to chat with the cook. "Ja, fish," he says, "we have plenty fish. Fish very good here, Sparks. We cook him good. Fry him, boil him, do most everything to him, and eat him." He laughs and continues, "Now you know what you get to eat for dinner each day. Fish, fish and more fish."

As he had predicted, our diet is fish after that, cooked in every way that the galley staff can think up. They are the best fish that I have ever tasted, fresh from the sea and well cooked. Most of the fish we catch are plaice, large oval flat fish, although several other kinds of fish are also caught on the fishing lines. Catching so many fish is a fisherman's dream come true.

We never did get ashore in Reykjavik. None of the crew were allowed shore leave because our stay was too short. Two or three days later we are ordered to Hvalfjord just along the coast from the city. This deep long fjord proves to be our rendezvous point with the Murmansk convoy.

Ships of many nationalities make up the Murmansk convoy. We tie up alongside a Russian ship for bunkering. I am surprised to see that there are women in its crew. I've never seen women working as crew members before. The long slow task of unloading the coal into ships' bunkers begins, and continues through the following days and nights. As one ship is bunkered, another takes its place.

"Ah, Sparks," says the Mate, coming up to me as I watched the coal being transferred into the ship tied up alongside us in the fjord. "How would you like to tally cargo for me?"

Put that way, how could I refuse. Besides it looks like fun and would help to pass the time. I just have to sit in the wheelhouse and count buckets of coal. The deck officers probably need a break anyway. "Sure," I reply, "but you'll have to show me how you want the loads counted and recorded."

"That's easy. Just make a vertical stroke for each load as

they come out of our hold. Do them in batches of four and for the fifth load just draw a diagonal line through the group of four, like this." He rapidly draws the tally mark for five buckets of coal.

"See," he continues, "there's nothing to it. Just record the tally marks on the tally record sheets on this clipboard." He adds my name to the roster of deck officers and their watch schedules. "Write the total number of loads in the cargo record book at the end of your watch, and sign the sheet."

That night, I take my turn tallying the cargo. Along with the deck officers, I mark each load as the ship's derricks swing the large buckets of coal across the deck, and into the coal bunkers of the hungry ships waiting impatiently alongside us.

The coal dust seeps into every nook and cranny, getting into our eyes, our ears and the pores of our skin. At the end of a four hour shift, it is fun no longer. I hand over to the next watch, and seek the relief of a hot shower. I lather myself slowly. The coal dust has penetrated to my scalp and I wash my hair vigorously. I'd love to have a bath, but there are only showers on the ship. Still the water cascading over my body is warm, and I feel clean again.

The night shift stands out in my mind. Beams from the large powerful lamps are directed over the holds and decks, and cast an eerie glow in the haze of coal dust. The shouts of men loading buckets in the hold is followed by the hiss and screech of the steam winches. The bosun or his mate directs the winch operators with hand signals, and the straining block and tackle of the derricks groan under the weight. The load of coal swings across the decks, and disappears in the coal bunker of the waiting ship.

Now empty of coal, the *Soborg* lies anchored in the fjord as the crew hoses down the holds. New orders are issued. We are to load barrels of fish destined for Aberdeen on the return voyage to Newcastle, the *Soborg*'s home port. Later that day, a tug towing the first lighters arrives alongside.

The ship hums with activity long into the night and the following day. Barrels bang against metal bulwarks as the steam winch operators bend to their task. The derricks creak beneath their loads. Once more, the holds are flooded with light at night as loading progresses nonstop. The night air is broken by intermittent shouted orders from the bosun to his men. The ship seems to sense the call of home, and is impatient of any delay.

As the loading draws to a close, I ask the Mate if anyone is going ashore. He smiles. "Ah, Sparks. Yes, this is your first voyage, isn't it!" He laughs as he looks towards the fishing village. It's the only sign of habitation in the lonely fjord. "Looks a barren place to me, but then, when I was your age, I was curious, too. There's a boat going off with the steward in the morning, though God knows what he'll find ashore. It's only a small fishing village with a single store. It probably supplies the whole community in this god forsaken place. If you want to catch the boat, then see the steward."

My first few steps on a foreign shore are taken in that lonely desolate fjord. As the Mate had said, it is a small fishing village. There are only a few stone houses and a general store. The whole village merges into the desolate landscape. A little group of men and women wander down the rough cobbled street, and look at me curiously and walk on without speaking. There is very little here, and I watch the steward go into the general store. For me, it is a milestone. I have walked on foreign soil and scrambled part way up the rocky slopes of the fjord behind the village. Doing so, marks an achievement, a beginning in my young life. Like the desolate fjord, the world stretches out before me.

The voyage back is violent. The weather turns sour again. Fighting our way in the teeth of exceptionally heavy seas and winds, we just about stand still. A five-thousand ton cargo ship capable of a top speed of about eight knots, the ship covers one mile in 24 hours and we lose the convoy. Eventually the weather and seas abate, visibility improves, and we find ourselves alone in an empty sea.

The Captain alters course for the Orkney Islands, and the safety of Kirkwall, the major port. Fortunately, there are no encounters with enemy forces on that run to safety. In those seas, the real threat comes from natural elements, not the U-boats. They have enough to do looking after themselves, and no energy to spare for hunting stray ships.

Our stay in Kirkwall is short, because there is a southbound convoy the following day sailing to the east coast ports in Scotland and England. We receive orders to join it as far as Aberdeen. The next morning, the *Soborg* leaves the anchorage and joins the other ships in a slow five-or-six knot coastal convoy.

The Orkney Islands are approximately 58 degrees north. On a cold November morning, we steam slowly out to sea in convoy formation, with naval escorts on each flank and rear. Heading south, we traverse seas that are typically storm-ridden at that time of year. The journey from Kirkwall is about one hundred and sixty miles, a short passage. The voyage is routine, apart from the continuing bad weather that moderates a little as the convoy slips down past Duncansby Head at the northern tip of Scotland.

The land mass of Scotland shelters us to a certain extent from the continual moving incoming weather fronts from the Atlantic, but the seas still run high. The wind howls down the North Sea whipping the wave tops and driving the rain in squalls, reducing visibility to two or three miles at the most. The weather masks our progress down the coast from enemy eyes, and we reach Aberdeen without incident the following morning.

Turning off from the convoy, we approach the anchorage and take on the pilot. We spend several days there tied up alongside the dock in the small Scottish port, as the crew and dock workers unload the barrels of fish. I take the opportunity to roam the quaint streets of this historic town, and pass the statue of Robbie Burns, the poet, in one of the many squares. While I wander the streets, the Chief, like many other seamen, roams the public houses.

I never did get on well with the Chief Sparks, but it is not until we reach Aberdeen that I realize the Chief is not a man to admire. Early one morning, I find him stretched out in a drunken stupor on the forehatch cover, laying fully clothed in a pool of his own vomit and excrement. I tell the night watch in case the Chief needs help, but the watchman only laughs and says, "Yes, I've seen him. The drunken bastard's all right. He'll sleep it off like he always does."

Apparently, the Chief is well known for his drinking habits. So much for respecting one's elders. I had learned another lesson:

elders must first earn that respect. After that episode, I lost any respect that I may have had for that particular officer.

In Aberdeen, our last port of call, I receive my first mail from home. There are several letters in fact, and I read them avidly. I know from my mother's letters and from the radio broadcasts that the nightly bombing raids on London are as busy as ever. These reports bring back memories of the London blitz, and I have a new urge to get home and see my mother and family.

My father died nearly three months ago, and I wonder how mother is adjusting to his death. In the seclusion of my cabin, my thoughts turn back to the old farm house. Although located in the green belt, it had been damaged by bombs in 1941.

Ludgate House, on the edge of Epping Forest

We leave Aberdeen on the morning of November 21, and are once more at sea. Our final destination is Newcastle. The mouth of the Tyne River lies roughly one hundred and eighty miles south and the port facilities of Newcastle are four or five miles up river. The Commodore ship signals a change of course. With ships on either flank and ahead of us, we churn through rough to moderate seas at the set convoy speed of five knots on a new southerly course.

As I come out on deck off watch later that day, the Mate pauses to talk to me. "Well, young fellah, this is the last leg of the voyage. We'll soon be in Newcastle now, and I guess you will be glad to get home. The first time at sea is always the hardest, but you're doing fine. You'll make out okay."

I smile at him, because these were the first words of encouragement that I have heard on the voyage. "Yes, I'll be glad to get home and see my family again. It's the first time I've been away, and it seems a long time."

We chat awhile before he turns away and continues on to his cabin. A warm feeling sweeps over me as I watch his departing figure, because he cares. The Mate is not remote like the Captain. He's human and probably remembers when he, too, was my age at sea. He is an older man, well in his forties and probably went to sea when he was fourteen or fifteen years of age.

The radio watch periods slip by slowly in an endless sequence of four hours on and four off. Each watch sees the ship getting closer to the Tyne, a sustaining thought for my tired brain. Seasickness and the stifling damp smell of the steam radiator, together with the radio static, are taking their toll. I need to get off this ship and rest my nerves for awhile. I am looking forward to home leave, even if it is just two weeks. Maybe my next ship will be a real deep sea vessel with a three-man watch system. I can but hope.

The coastal traffic on the high frequency bands (HF) is heavy as we edge down the coast, indicating enemy action from aircraft and E-boats in the Channel and the coastline of southern England. Our convoy is lucky. Unlike the English Channel and the south east coast of England at this time, our area is quiet. Hidden under the cover of thick low-lying clouds and squalls intermingled with fog banks, we slip through to reach the Tyne without incident. I experience a sense of relief when I finally see the Tyne estuary ahead.

Flags of different colours and patterns fly from the mast of the commodore ship as we prepare to leave the convoy. Answering flags are raised from our bridge and the bridges of other vessels also destined for the Tyne. Their import is a mystery to me but fortunately the deck officers understand them. We drop out of our convoy stations and in the company of several other vessels turn and head for the river. Now depleted in numbers, the convoy sails on south towards the attacking enemy forces. I watch them go with mixed feelings.

The end of the Jacob's ladder dangles a foot or two above the deck of the pilot's launch that draws up alongside. The pilot waits his chance in the rolling sea, then seizes the swaying rope ladder and clambers up. He swings his leg over the bulwarks, drops onto the deck, and walks swiftly to the bridge.

The engine room telegraph swings to half speed ahead as

the pilot takes over the ship, navigating it safely through the marked channels of the Tyne Estuary. We are back and even the drab docks and flat countryside look good to me as we slowly pass up the broad dirty river.

The *Soborg* returns to the Tyne on the morning of November 22, 1942, and docks alongside in the early afternoon . The Ministry of War Transport officials come aboard the next morning to sign the crew off the ship's articles. I wait until the crush of crew members has thinned out before approaching the little group of men at the table with the Captain. He thanks me and wishes me well as I sign off and make my way out of the wardroom.

In ones and twos the crewmen make their way to the dock gates. I glance back to see the collier as I, too, walk through the docks. She looks much like she did when I first saw her in the London docks. She is dirty and grey, with paint peeling off her and has the same dejected and forlorn air. The *Soborg* still retains a place in my heart despite all the stress and frustrations that I experienced aboard her. John Masefield's poem will always remind me of the collier with her "cargo of Tyne coal".

I catch the London train that afternoon and settle back in the compartment, as we chug and clank down the track to war-torn London. The nightly bombers will welcome me but so will my family and that is where I want to be.

* * *

Of Home and War 1939-1941

My mother came from the north, from the moors of Yorkshire, from Easingwold a small village not far from the ancient fortified City of York. York was rich in history and, of course, bloodshed. Her father hailed from a long line of prosperous millers in Yorkshire who went by the name of Dunn.

Grandfather Dunn was a stocky, white-haired man with a little Hitler moustache, rather awe-inspiring and autocratic. Mother claimed that he had attended a famous school in the North of England called the Bluecoat School, but then my mother used to tell all sorts of stories.

According to my mother, my grandfather was a bit of a tyrant, and there was not too much love lost between the two of them. I can still remember the day that she handed him my father's double-barreled shotgun and told him to go out into the forest and shoot himself. He did not take her up on the offer, but it was the last time that I remember seeing him.

I never knew my Grandma Dunn because she left England

when I was very young. She went to Australia to visit my Aunt Eileen. In those days the voyage via the Suez Canal would take approximately four or five weeks. She became ill during the voyage, however, and died on October 26, 1926, a month after her arrival. She was buried just outside Freemantle near Perth, and my uncle used a hot poker to inscribe her name and dates on the plank that served as her tombstone. It's hard to decipher the year of her birth on the blurry photograph that I have, and the cemetery was destroyed by wildfires many years ago.

According to my mother, Grandma Dunn was an orphan born in County Cork and raised in a convent. Grandfather Dunn had talked her out of becoming a nun and married her. Some family members say they met in Paris. Others believe that they met on a passenger ship: my grandfather was evidently the ship's purser and my grandmother was a governess in charge of a passenger's children. Although various family members have tried, no one has found out about her past.

My father's family was from Essex in the south country. Country folk, they made their living either by working in the forest or for local landowners. They raised a few vegetables from their small patch of land, supplemented by trapping rabbits and possibly a little poaching. My Grandfather Mumford was a tall man with a Kaiser moustache, slightly stooped with age, while Grandma was grey and plump. This vague description could fit anyone's grandparents from that era, but I have no pictures and can recall no other details.

The Mumfords were tall. My father was well over six foot, as was his brother Sam. His parents had adopted my Uncle George as an infant. He was about six years older than I was. He was also tall, but had a small head.

My parents had met in Yorkshire, at the end of the First World War. They married and my father brought my mother south where they rented a cottage, located on the edge of Epping Forest not far from Chingford. The forest was about twelve miles long and four miles wide, and stretched from Chingford to Epping, and a popular destination for daytrippers from London.

Known as Ludgate House, the building was about five hundred years old, and was reputed to be a hunting lodge and part of a large estate. The front of the house jutted into Epping Forest. According to my Uncle George, the old house had been deserted for several years before my parents moved there in 1923, and alders were growing in the dirt floor of the main room.

I took the old farmhouse for granted. It stood isolated, a countryman's half mile from the nearest road. As children, we used to walk a well-known path through the forest and fields from my grandparents' house to my childhood home.

Ludgate House, Epping Forest.

The house was L-shaped, and a large barn, used to store camp equipment, tools and a hundred and one things, formed the long leg of the L. In those days, there was no electricity or water laid on to the house, and my father fetched drinking water from a well near the Owl Inn, the local pub that was about a mile away. He trundled over the fields with a curved, wooden yoke resting on his shoulders. Two three-gallon containers fastened by rope and hooks hung from either side of the yoke. This strange looking contraption was extremely useful for carrying water over long distances.

From the barn, the back door opened into the scullery, where there was a washstand and sink. The drinking water container was next to the sink, and there were usually a couple of rabbits hanging by the back door. There was also a small larder off the scullery where my mother kept our food. The laundry was done in the scullery, and rainwater was used to wash clothes and for bathing. Large catchment butts were strategically placed around the house to collect water from the drain pipes off the roof.

Saturday night was bath night, and buckets of rainwater were heated on the Dover stove in the main room. The warm water was poured into an old tin bath, and family members took turns bathing in front of the Dover stove.

The main room was where we lived, ate, and cooked our food. It opened off the scullery and was about 16 feet square. The kitchen range, known as a Dover stove, was used to cook, bake, and heat our bath water.

There were four children in our family. My sister Eileen was the oldest, nearly three years older than I was, while my two brothers were younger. Peter was five years younger, while Douglas was seven years younger.

I can still clearly remember the outside drafty wooden outhouse, and how bitterly cold it was sitting there in winter. Odd shaped pieces of newspaper, cut ready for use as toilet paper, were impaled on a spike in the wall. Often we would read cartoons from the Daily Mirror. One of my favorites was Jane, a comic-strip character who seemed to have an ingenious capacity for losing her clothes at the slightest provocation. Another favorite was a cartoon about Garth, a superman with mystic capabilities, and Professor Luminere.

Although she was not religious, my mother raised us as Catholics. She was a very practical person, and I suspect that she did this in order to get reduced fees at the Catholic schools and colleges in our neighbourhood.

In fact, my mother embraced any crusade or philosophy that happened to pass her way. She was an ardent communist, and eventually became a card-carrying party member. In truth, she was an entrepreneur and rented the communists land for the large, organized rallies held prior to the Second World War, and then sold them refreshments. These meetings were all well-supervised by the local constabulary and punctuated with speeches and campaign buttons.

I have soft, fuzzy memories of those happy and sad days of my childhood, of my fears, superstitions and hopes. Odd memories of events stick in my mind from those days before the war began. I can remember running away on my first day at school. I also made a myriad of little dens and hideouts in the forest surrounding my home. Perhaps that is why I later made so many tree houses for my sons when they were young.

Looking back into the past can be a double-edged sword, but fortunately I seem to remember mostly the better things in my childhood. I remember the awe I felt when I saw an airship in 1936, the only one that I have seen. The airship was huge and seemed to fill the skies over London as it moved slowly overhead. This was the last flight of the R101. Not long after I saw it– in fact, only a few hours later-- it crashed in France.

The more one thinks about one's childhood, the more that comes to mind. I guess that doing stupid things is all part of growing up. One time, I tied a string to a dart to avoid walking to the dart board to pull the dart out. Unfortunately, the string was too short, and *Wow*, the dart stuck in my ankle. My friends and I had fights with our airguns in the forest. Once, I fell from a tree and stripped the flesh from my smallest finger.

In the depth of winter, ice sometimes formed thick and hard on Connaught Waters, and we went skating. Sometimes we had snowball fights. I recall walking with my parents through the forest in spring and summer, wandering through beech trees, silver birch, oak and elm trees. In spring, new saplings sprouted through a tangle of ferns and undergrowth. , interspersed with the occasional forest clearings lush with fresh green grass, cropped short by wild deer and other animals.

My memories of Epping Forest where I was brought up are warm and in some ways nostalgic. The forest was beautiful, an ideal place for a young child to fulfill his fantasies. The forest changed in mood and appearance as the seasons followed, one after another. The stark, cold, winter days made the forest forbidding because the trees were stripped of their leaves and the branches were bare. But the snow added a majesty to the forest.

As spring approached, the trees came alive again and everything seemed right with the world. I can still feel the wonder and delight at being alive, as I walked through the forest on my way to and from school. I loved the forest with its scents and blossoms, its solitude and mystery.

Then came 1938. Munich and Chamberlain. Peace in our time was the familiar slogan. Uniforms appeared in the streets and I was in the Air Defence Cadet Corps. Life was full of excitement, and we wanted to be a part of it. Drills and courses in aircraft design, the principles of flight and many other courses like armaments occupied our evenings and weekends. We felt like men.

* * *

On September 9, 1939, war was declared. I can remember that day as clearly as if it had happened yesterday although it happened over sixty years ago. The Prime Minister announced that Britain was at war with Germany. The voice on the radio stopped, and a band played miliary music.

My parents had been listening to the old radio in the dining room. My father sat staring silently into space with tears streaming down his face. Mother rubbed his neck and shoulder blades, trying to ease his pain, as she murmured softly, "It's all right, Jim, it's all right, love."

Both my parents seemed to have suddenly aged. Tears stung my eyes as I watched. I was fourteen, but I had never seen either of my parents cry before.

"Dad, don't cry." I pleaded as I ran to him and tried to hug him, but my mother gently pulled me away.

"No, love," she said, "leave your father be. He'll be all right in a minute. It's the bloody war again." Her eyes were red and watery as she, too, struggled to hold back tears.

Suddenly, all the air raid sirens shrilled sharply. The sound seemed to last for several minutes. It turned out to be a false alarm, one of many to occur in southern England.

I never went back to school that fall because there was no school to go to. All the schools in the London area were evacuated to the country. St. Egbert's College, the Catholic college of the Order of the Sacred Heart which I attended as a day scholar, was evacuated to Rugby in the Midlands of England.

My parents could not afford to send me there as a boarder. I was fourteen at the time, and like all young teenagers impressed with the excitement and jingoism of war and soldiers, marching bands and the hoopla that war generates. I would learn, of course, but at that time in history, life took on another more exciting turn that consumed my thoughts.

* * *

The Ministry of Defense and the RAF arranged to have the Air Defence Cadet Corps work on military airfields. Similar to the Hitler Youth movement, except for the Nazi ideals, we cadets assisted and learned a wide variety of jobs in the armed forces. I had joined the air cadets in 1937 when I was twelve. Now at fourteen I felt a man, proud of my uniform and our unit.

At last I was close to my dream of touching and climbing

up on a hurricane fighter, and, as I walked with Taffy, an RAF armourer in a fighter bomber squadron in the Northweald airfield, I felt ten feet tall.

We used a small hand-pulled tender to carry the belts of machine gun bullets to the hurricane fighter aircraft, because the belts of bullets in their feeder containers were very heavy. The belts were patterned with a combination of armour piercing, incendiary and standard bullets, arranged alternately.

I can remember sitting on the wings with Taffy loading the belts into the breach blocks of the Browning machine guns. There were eight of these guns in the wings of each fighter aircraft and their firepower was arranged to converge at three or four hundred feet ahead of the aircraft to inflict maximum damage.

I wonder what Taffy thought about all those young air cadets swarming over the airfield, climbing into aircraft, and learning various jobs that included dismantling machine guns from the aircraft and then checking and cleaning them under the supervision of a leading aircraftsman and armourers like my friend Taffy.

Gordon in his air cadet uniform, 1937

To the air cadets, it was the proudest moment of our young lives just to be there. Taffy was, of course, a Welshman. Who has ever heard of anyone by that name not coming from Wales?

Digging back into my memory, I can still remember him. He was a short man with dark hair, a kind man who took us in his stride. Even when one of the cadets accidentally fired a round from a machine gun while cleaning it in the workshop, Taffy stayed calm. Luckily no one was hurt when the gun went off. Mind you, it must have caused some concern for I remember the incident after all these long and eventful years.

Day after day, all through that autumn of 1939 and on into the winter, when the rain turned into snow and the frost laid a hard covering over the ground, we boys boarded the RAF transport truck at seven each morning bound for Northweald airfield and returned in the early evening to our local headquarters.

The airfield was a fighter and fighter/bomber station in the green belt surrounding London, located just north east of the city. The aircraft were Hurricane fighters and Beaufighter bombers. They played a large part in the Battle of Britain fought in the air mainly over London and the south country.

Towards the spring of 1940, all air cadets were withdrawn from the RAF air fields throughout Britain, due to the increasing threat of attack by enemy aircraft on these airfields. Our training schedules were reorganized and scheduled courses on aerodynamics and other subjects were taught at the headquarters building as well, of course, military drills and parades. A little later the Air Defence Cadet Corp. was renamed, and became one of the main avenues for recruiting youths into the RAF.

I turned fifteen in April 1940, and the spring and summer were beautiful that year. The weather was hot and sunny with clear blue skies for months. That summer stands out clearly in my mind. The forest was no longer peaceful, because with war came bombs to blast and crater and shatter its solitude. Bombers dropped their loads prematurely on the outer edge of London's antiaircraft defences, and parachute mines floated down at the whim of the prevailing winds. Even our old house suffered. Bomb blasts blew in doors and windows, and ripped tiles off the roof.

The things that I remember about that period of the war are the dramatic ones, and the bombing of Northweald air base was certainly one of those moments. I was about a mile or so from the airfield on my bicycle when I saw a squadron of German bombers flying in tight formation through a thick blanket of antiaircraft flak. Not one of the aircraft deviated from the formation, but bore on steadily. They dropped their bombs and then swung away in a wide arc to return to Germany.

The sky was cloudless, and high-flying aircraft used their vapour trails to draw swastikas in the upper atmosphere. A massive attack force of five hundred bombers filled the sky, almost blacking out the sun.

Barrage balloons prevented aircraft from flying at low altitudes over London to bomb specific targets or machine gun people on the ground. In the countryside around London, bursting shells from antiaircraft guns dotted the sky with little black balls of smoke. I caught a momentary glimpse of a German Messerschmidt hedgehopping in an effort to shake off an attack by a Hurricane fighter.

The daylight bombing raids on London had ended in late September 1941 and were followed by night bombing raids. During the night raids, bombs and parachute mines were a nightly occurrence and I used to watch the panorama of war from the front porch of our house on the outskirts of London. When the antiaircraft battery located on a hill a short distance from our farm house opened up with their 4.75-inch guns, the house shook with the noise and concussion of the gun blasts.

I happened to be outside during one of these raids when I heard the shriek of falling bombs, and saw a whole stick of bombs burst in an ever-approaching line that was heading for our house. Shouting a warning to my parents and family, I threw myself flat as I watched the flashes and felt the concussions of the bombs coming towards us.

Suddenly, the little bluebell wood, where my sister and I had once played, was lit by a brilliant flash of light when a high explosive bomb exploded, ripping and rendering the trees. This was immediately followed by sharp shrieks from the flukes as an oil-filled incendiary bomb hit the path, exploding just behind the house. Barely a hundred yards away, the burning oil ignited the elm trees that flared like torches, filling the night sky with light.

* * *

Down to the Sea in Ships

By 1941, the great daylight air raids were over, and Dunkirk was a thing of the past. The navy stood between German-occupied Europe and the British Isles, the Home Guard had been formed and trenches and other defensive earth works were dug all over the countryside. The nightly air raids continued with monotonous regularity.

That April I turned sixteen, and feared the war would be over before I turned eighteen, and old enough to enlist. The urge to join the forces and get into the war ate at my soul.

In a way it was my father who inadvertently gave me the opportunity when he found me a job in the Enfield Small Arms factory as an apprentice. Restless, after three months I gave in my notice and set off for central London in search of a shipping company where I could sign an apprenticeship as a navigational officer. They took apprentices as young as fourteen.

I found work in the City of London as an office boy with a tea brokerage firm, Brash Bros. Ltd., 13 Rood Lane, Fenchurch Street, while I canvassed the shipping companies for an apprenticeship. Working in an office as an office boy was a new experience and I used to think that the young ladies in the office were goddesses. I travelled by train each day to Liverpool Street Station, a dull dirty smoke grimed station that served the north-eastern suburbs of London.

Miss Irving was a tall woman, middle-aged with graying dark hair, who held a senior post in the Company. She was one of two tea tasters; the other tea taster was a man. I always remember her standing by a long table filled with cups of different blends of tea. She would wander down the table, take a small sip out of each cup, swill it around in her mouth, then spit it out in a handy receptacle and make an odd note or two.

Eddy was the junior clerk and had short black curly hair, while Muriel, another of the office clerical staff, had long fair hair and a friendly manner. I looked with awe upon Muriel. To me, she seemed so sophisticated and attractive.

In my lunch hour I used to haunt the shipping offices in the City looking for a job as an apprentice aboard ship in the Merchant Navy. Eventually I found a position with Andrew Weir & Company, whose ships traded along the West Coast of Africa.

An elderly man—he was at least thirty—leaned back in his chair behind an imposing office desk and looked up at me over the top of his glasses. He smiled and said, "Well Mr. Mumford, you have passed your medical. The company have accepted you as an apprentice deck officer. As you are only sixteen, you will have to get your parents' permission before we can employ you."

Handing me some papers, he stood up and continued, "Get them to sign these documents and return them to this office as soon as possible." He laughed at my evident pleasure and wished me well. I hurried from the office, my mind in a whirl. There was now only one further hurdle to pass, my mother.

Later that evening, I picked an opportune moment when Mother was alone in the kitchen. "Mum," I spoke hesitantly, "I've found a job with Andrew Weir, a shipping company, as an apprentice deck officer, but I need your permission. They gave me these documents for you to sign." I handed her the papers.

She sat down at the kitchen table, her worn face drawn in a frown of concentration as she read the terms of the apprenticeship. She glanced up at me sitting anxiously opposite her, my arms resting on the table. She knew how much this meant to me, her eldest son, but she shook her head.

"No, Gordon, they 're sinking too many ships right now, I can't let you go.

My heart sank at her words, but then she continued with a half smile on her face. "However, if you really want to go to sea that much, you can go as a Radio Officer like your Uncle Jack."

Jack Law was an old family friend who used to visit my mother and father. Ever since I can remember, Jack had been upheld as a role model.

"Now, listen," she said, as she saw the relief in my features, "I'll have to talk to your father about this, but don't worry, it'll be all right, son."

I was delighted. I knew the battle was won. Dad didn't have a chance. Mother was the boss in our family, at least that's how we kids saw it, because our dad simply seemed to fade into the woodwork. Mother was the dominant character in our lives.

As a result of the ensuing family discussions, I enrolled in the Holloway Radio College in September, 1941. The timing was perfect. I wanted desperately to get to sea before the war ended, and now I could begin my training for a career that I had longed for. Like most youths of my generation, I lived in dreams and in the excitement of action. I did not know what lay ahead; perhaps it was just as well.

Every day through the winter of 1941/42, I bicycled about ten or twelve miles to the Radio College and back again . The days were cold and wet with occasional snow and freezing weather that winter, and cycling across the slippery roads, thick with traffic, was no easy matter for anyone, let alone a sixteen-year-old youth.

May 1942, graduation day. Our little group of five were elated. Clutching our seagoing certificates, we exchanged addresses and plans. The five of us had stuck together as a group through our training and were fast friends. Only two of us would survive the war. Don Cummings, who lived not far from my home, died in the Atlantic a year later. Chris and Bob were killed, one off the coast of Sicily when his ship was sunk during the Sicilian landings, and the other blown up on a tanker off the Algerian coast. I would be torpedoed in the Atlantic in December 1942 and sunk by a mine in the North Sea on Christmas Eve, 1944. Eric came through without a scratch; he was fortunate.

Shortly after I graduated, my father died; he was forty-five years old. I was now the breadwinner for the family. I joined Siemens Bros. Ltd. as a marine Radio Officer in August 1942. Four weeks later I was at sea on the *Soborg*, an old Danish collier.

* * *

The Talisman

Brother Isidore's voice was harsh and grating. "Your mother is a communist, a card-carrying communist." Feeling a flush of shame, I started to protest, but the principal held up his hand, palm out, silencing me. "Let me finish," he said. "Because your mother is a godless woman, it didn't seem right to give you a sacred heart three years ago when you left the college." His words shocked and bewildered me.

Although the day was cold, the sun was shining fitfully through thin scattered clouds. On impulse, I had decided to visit St. Egbert's College, my old school, a Catholic college run by the Order of the Sacred Heart. I had put on my uniform. I took pride in wearing it because it made me feel like a man, but Brother Isidore had made me feel like a child.

To me, these monks in their long, brown gowns were like gods and I had always been overawed by them. The God they worshipped was not a loving God, however, and their students learned to fear God and the wrath of their teachers. Although I had been away from the college for more than three years, they still had a powerful hold over me.

Brother Isidore paused a moment, and his eyes glittered

behind the small rimless spectacles. "After your recent experience in the Atlantic, I feel that I should probably give you the talisman now. You could have lost your life when your ship was torpedoed." He was silent a moment, his eyes downcast, hands fumbling in the folds of his gown. "Four ships sunk before yours! How many more?" He muttered the words to himself, leaving the sentence unfinished.

Glancing up, he handed a small blue box to me. "Here, take this sacred talisman," he said. "The bleeding heart is the symbol of our Order, and may protect you."

Perhaps I should have refused to take the talisman, but I was unable to think clearly. The box felt warm, and fitted snugly in the palm of my hand. As we stood together in the entrance hall, I lifted the lid. A small red heart, its edges defined by a fine gold thread, lay against the blue and white lining of the case. I touched the velvety material of the talisman gently with my finger.

"Take care of it," Brother Isidore said as we parted. "It may save your life."

Crossing the school quadrangle, I glanced back at him. His bulky body was cloaked in a dark brown, ankle-length gown which almost touched the floor; a cowl hung down the back of his neck, and a large red cross was sewn down the front of his gown. His heavy jowls partly hid the stiff white clerical collar.

* * *

By the time I got home, it was dark. Dusk falls early in midwinter and I had dallied at the school talking to old acquaintances. The back door creaked when I opened it and entered the scullery of the farmhouse. Mother was preparing supper and

I smiled as I heard her cussing out the kids. She'd not changed in that respect. "That you, Gordon?" she called. "Dinner's almost ready."

Later in the evening, I sat alone by the log fire in the front room, which now served as my bedroom. Turning the talisman over in my hands, I pondered Brother Isidore's remarks. I smiled as I remembered the communist rallies held in our field before the war. Sure, Mother was a communist. Why shouldn't she be? She was the businesswoman of the family and made a profit from all those rallies. Her tea, cakes, sandwiches, and ice cream sold like hotcakes.

Mother was educated, and managed the money end of our small camping and catering business. True, it was primarily a weekend business, but it was a profitable one, especially during the summer months before the war. Dad had worked as a landscape gardener for seven shillings a day and got a bit more on weekends as the golf pro on a nearby golf course.

The flames shot higher as I fed the fire with the logs I'd cut and split. Since my childhood, flames had always fascinated me. My father used to sit on this fireside seat where I now sit, poking the logs into flame. I asked myself, *Why did he have to die? There was so much that I would have liked to ask him.*

The *Scottish Heather* had been torpedoed two months ago, and had struggled back to the Clyde. I had been on survivor's leave since then, and now I was restless, ready to go back to the companionship of the sea. The telegram from my company was vague; it just gave the ship's name, *Empire Harmony*, and routing instructions to the docks in West Hartlepool. All I knew was that she was an ocean-going vessel, not a coaster, but not getting much information was normal in wartime. A smile lit my face. I'd been promoted from third to second radio officer, and now there was a second wavy gold band on my uniform sleeve.

My suitcases lay open on the bed as I packed. Hesitating, I picked up the talisman and examined it. The small heart was an enigma, a symbol of superstition, like a lucky charm to ward off evil spirits. For whatever reason, I could not abandon it of my own volition. The influence of the Roman Catholic faith of my childhood was still too strong. Deep in me lay a superstitious fear, a certainty that if I destroyed the talisman, a greater harm would come.

In late March, I was on my way. A feeling of excitement edged with uncertainty surged through me. As I watched the passing countryside flash past the train's windows, I wondered when I'd see London and my family again. The talisman had been safely stowed in one of the suitcases. As I sat in the railroad carriage of the north bound train, my cases were in the luggage rack overhead. Soon London along with its nightclubs and theatres was behind me.

* * *

After eighteen months' service on the *Empire Harmony* in the Mediterranean, I was sent back to England. Not long afterwards, I transferred again. On Wednesday, November 22, 1944, I went to the Tilbury docks where I signed on the *Empire Path* as the second radio officer.

The *Empire Path* was loaded with war materiel, and we sailed in late November for the Scheldt Estuary in Southern

Holland, en route to Antwerp. After navigating the riverways, we arrived on November 29. Antwerp was the final staging point for the invasion of Germany. While the ship was being unloaded, the port was under almost constant attack. On December 24, we started on our homeward journey.

When we reached the estuary of the Scheldt in mid-afternoon, the tension seemed to lift. Christmas Eve dinner was being prepared. Suddenly, the festive mood was shattered and our world turned upside down. A tremendous explosion tore the ship almost in half, like a giant fist thrusting through the bottom of the ship.

J. Gordon Mumford in Belgium, November 1994
for 50th Anniversary of Battle of the Scheldt

We were in the midst of the convoy, and were quickly picked up by the naval escort. We watched from the deck of a Motor Torpedo Boat (MTB) as the ship sank, wallowing in the sea with her back broken.

Many years later, I discovered that the *Empire Path* had been sunk right on top of a World War I wreck, the *SS Boscobe*, raising her superstructure to just under the surface of the sea. A few months later, the *SS Emeraude* hit the superstructure of the *Empire Path*, and ripped out the bottom of her hull. Ironically, the mine that sank the *Empire Path* had sunk two ships.

These three wrecks were now a shipping hazard, and at the end of the war they were blown up. The remnants of these wrecks and the talisman are now scattered over the seabed of the North Sea.

The talisman is gone. Perhaps it did save my life. It did not die on the *Empire Harmony* in the Mediterranean in Operation Torch, or in the Italian and North African campaigns, but eighteen months later on Christmas Eve, 1944, in the Battle of the Scheldt Estuary. The talisman now lies on the seabed off the coast of Holland, along with the *Empire Path* and her dead, all left to rot in a watery grave. But I survived.

* * *

The Canoe and the Bangkok River

It's two or three weeks since the *Luling*, a small tanker of 1700 tons, arrived in Klong Toi, a small naval base that the Navy took over from the Japanese. The small docks are located up the Bangkok River, often referred to in Siamese as Chao Phraya.

We are part of the naval cleanup operations in Siam (Thailand) in 1945. Our shallow draft tanker can traverse the sandbar at the mouth of the Bangkok River at high tide, something that naval vessels cannot do. Our mission is to bunker the ships downriver at the anchorage of Lem Sa Muk with fresh water. For us, the routine of three days in port and three days in the anchorage is ideal.

The Chief Engineer is an unusual person. His first name is Reg, but I just call him *Chief*. Somehow, calling him that seems more natural. He is in his early thirties, well built and handsome, in a hard-faced way.

Shortly after our arrival, the Chief and I bought a canoe. A mutual interest in canoeing and a need to get away from the confines of the ship have brought us together. Painted eggshell blue, the canoe is about fifteen feet long with a sleek, strong wooden frame covered with thin, lightweight wood. To us, she is a thing of beauty that glides effortlessly down the river.

We have free time today, and soon the small tanker is astern, and the docks are hidden by the dense foliage growing along the river banks. Hugging the southern bank, we round a bend in the river and find ourselves alone. It's peaceful as we drift slowly past the reeds and bush of the forest. Occasionally, we hear monkeys swinging through the branches somewhere above us and the soft hum of insects. A bird call breaks the silence, warning others of our approach.

I am lazily watching the thick reeds that hide the banks, when I notice a small break in the seemingly impenetrable wall of tall reeds. "Look," I call out, "there's a small opening in the papyrus. It must be a waterway."

"You're right," the Chief says, steering towards the gap. "Let's see where it goes."

Ever since I first noticed the small native boats suddenly emerge, as if by magic, from the tall reeds and rushes lining the river, I've wanted to explore these hidden passages. Now we are on the verge of discovering these secret waterways.

We pass close to the far bank, and I catch hold of a papyrus stalk, but it moves with me. Startled, I look at it and discover that a clump of papyrus has broken off. "Christ," I exclaim, "that's strange. The roots don't go down to the river bed. The papyrus just seem to be floating on the surface."

The Chief laughs and replies, "Be careful of that stuff. If you fall, the papyrus will sink with you. If you get tangled in the reeds, you'll drown."

Grimacing at the thought, I nod. "Yeah, it's either the pollution or the reeds. If one of them doesn't get you, then the other will. Not a very nice thought."

I've been reminded of something the Second Mate said when I asked if he wanted to go canoeing. "That bloody water is polluted to hell. Christ, man, we're only eight miles downstream from Bangkok. There's no sewage treatment system there. Never has been. All the city's raw sewage flows directly into the river. You'll get typhoid or some bloody disease if you fall into that lot.." He emphasized his words by pointing to the river. "Just look at it." The water flowing past the ship was a thick, unappetizing brown colour, either from erosion of top soil or unsavoury refuse.

When I tell the Chief this, he only grins. "Well," he says, "It's not quite as bad as that, Sparks. Anyway, I don't plan on getting sunk or swimming in this muck."

* * *

We edge the bow of the canoe into the channel, and slip through the papyrus. It towers six or eight feet above us. The long slender triangular stalks are topped with tufts of fine fronds that radiate out from the tips.

Our paddles stir the still, reed-locked waters, and send little ripples across the surface to disturb the water lilies lining the cleared waterway. They bob up and down in the ripples, their large green leaves contrasting vividly with their pink and red flowers as they move gently with the passing of our canoe.

Gradually the papyrus gives way to swampy ground and

forest. The papyrus no longer floats but the roots burrow into the soft mud of the swamp. "Bloody insects," the Chief growls. He slaps the back of his neck where a swamp mosquito has found a sensitive spot.

The day is hot and humid. Wiping the sweat off my face with the back of my hand, I brush away the tiny irritating sweat bees that cluster near my eyes. The channel winds through a swamp land of dead fallen trees and bush. The canoe slips past mud flats studded with tangled thornbush. The smell of brackish water and rotting vegetation assail our nostrils. Strange stunted trees, misshapen and grotesque, stand like sentinels in the swamp. Fungus clings to the tree trunks, and the bark and branches are festooned with lichen. The trees are perched precariously on roots resembling legs, and look like they are trying hard not to get their feet wet.

Firmer land and palm trees, interspersed with thick rich vegetation, welcome us when we leave the swamp and enter the forest. The Chief sighs with relief. "Jeez, it's good to be out of there," he mutters. "That stinking place was getting on my nerves."

I nod my agreement. There had been times that I wondered whether we'd missed the channel. *It would have been a hell of a place to get lost.*

The banks of the waterway are now clearly defined as it snakes deep into the low lying land. Hard packed earth, trees and palms together with luxurious tropical forest growth replace the swamp. We pause momentarily beneath the shade of a tree on the edge of the channel.

I steady the canoe with one hand gripping the root of a tree protruding from the bank. The bark is smooth to the touch. An ant runs across my fingers and scuttles away. There must be an ant nest nearby.

"This is great," I say. "It's so quiet here."

"Yes," he replies. "It's the reason I wanted the canoe. It's a relief just to get away from the ship." Half turning, he continues, "Look at the trunk of this tree! It looks hairy, doesn't it. Must be fibrous, something like a palm. Probably related to them, but the trunk is much thicker."

It's pleasant in the shade, and we sit awhile, taking in the scenery. It's time to move on. Reluctantly, the Chief picks up his paddle.

* * *

We are not alone. The bush has eyes. At first, we did not notice the little tree houses hidden in the foliage of trees and bushes. As we paddle down the waterway, a movement catches my eye. I have a brief glimpse of a woman's head and shoulders as she peers down from an opening in a tree house above us. The hut blends in with its surroundings, and only her movement has betrayed its presence.

"Look, up in the trees. There are tree houses," I say, pointing with my arm. The head disappears.

Startled, the Chief glances up, and sees for himself. The huts are built on a platform of thick bamboo poles, raised fifteen or twenty feet above the ground on stilts formed from tree trunks. We wave and tiny hands of children wave back as we sweep round another bend in the channel.

"Did you ever build tree houses when you were a kid?"

The Chief laughs and shakes his head. "No, I never had the opportunity, at least, as far as I can remember. I was brought up in a town." He hesitates and then continues, "Those huts remind me of Robinson Crusoe and man Friday, but you're from the country, aren't you?"

"Yes," I say wistfully, my eyes fixed on the paddle as I dip it into the water, and watch the little swirls trailing behind the dripping blade. "I used to build many tree houses, because our house was almost entirely surrounded by forest. We lived nearly a mile from the nearest road or house. I loved those little tree houses and dens that I made. I think some of my happiest moments as a kid were spent building tree houses and playing in them."

* * *

About midday, we come across a temple in a small clearing on the edge of the waterway. Stone steps cut into the bank lead up to the forecourt. A group of children play nearby while a man in long robes stands to one side watching them.

As we bear down on them from around a bend in the waterway, the suddenness and silence of our approach takes them by surprise. We paddle the canoe towards the small landing stage at the bottom of the steps, and, as our intent makes itself apparent to the children, they scatter and run.

The man comes slowly towards us, his hand held up in greeting. When the children see that he is approaching us, they stop running. They linger cautiously at the edge of the bush, peering at us. We see the man is old, with a wizened and wrinkled face.

A smile lightens his features as he speaks, probably welcoming us in his native tongue. Then in faltering English he adds,

"Hello," and beckons us ashore.

"Better take our lunch here," suggests the Chief, "It's about time anyway."

I nod briefly and bend to pick up the small packages from the bottom of the canoe that the steward had prepared for us. As I stand, the canoe rocks slightly, and I step out and follow the Chief up the short flight of stone steps.

The Chief and the old man are trying to communicate when I join them. "Where you from," says the old man, his brow wrinkling with concentration. He emphasizes each word with gestures.

"Ship," the Chief replies. "Ship, you know, all same canoe, but big." He pantomimes his words, first pointing at the canoe, then stretching his arms far apart. In desperation he says, "Ship, Klong Toi."

Comprehension dawns in the old man's eyes. "You," he stabs a finger at the Chief's chest, "Ship, Klong Toi." The Chief nods his head vigorously. The old man continues. "Ah, yes. Japanese before, now English. You English."

The Chief grins with relief and says, "Yes, we're English." He turns to one side and laughs as he says to me, "It's a good thing that we're not Scotsmen, Sparks, or we'd be here all night trying to explain the difference."

Using signs, the old man asks us if we would like to eat. We thank him and offer to share our food with him. He speaks to the children who bring coconuts in their husks and glasses. The old man lops the tops off the husks with a machete to extract the nuts. Then he cracks the coconuts open and pours the liquid into the glasses set before us.

We sit on a log under the shade of a palm tree on the edge of the clearing. The temple, a solid looking wooden structure with a high carved entrance and whitewashed adobe walls, is partly obscured by the trees and flowering bushes that flank it. Pink and red bougainvillea grow in profusion, interspersed by the fleshy-leafed frangipani and other tropical plants, lending an air of peace and beauty to the temple and its surrounds.

We share our lunch with the children. With the unabashed curiosity of the very young, they stare up at us with wide-open eyes, solemnly inspecting our features. Giggling, they turn and chatter excitedly amongst themselves. A small girl with black hair and large brown eyes looks up at us, her finger in her mouth. I doubt whether she's ever seen a white man in her young life, because she stares at us with wonder in her eyes. She looks at the old man sitting next to us for confirmation and reassurance as we offer her one of our sandwiches. He smiles and nods so she gingerly takes the food and tastes it.

The old man's face reflects his feelings as he watches the children. He smiles as he chatters to them in their own language. Unlike the children, he is more reserved and declines to eat our offerings.

The sun is high overhead when we leave the clearing and temple. It's time for us to start our homeward journey to the ship. Our little group has gradually swollen in numbers as nearby villagers come to see the white men, clad in shirts and shorts, who travel in a small canoe. We cast off and paddle slowly up the waterway and through the swamp and papyrus to the river, leaving behind a small crowd of villagers waving and talking excitedly amongst themselves. Our visit has caused quite a commotion in their lives.

As we paddle against the current, steering close to the south bank, the *Luling*, still docked alongside, hoves into sight. Although we may often curse her, the ship represents something solid and familiar in our lives. We are home again, back in our own environment.

* * *

The weeks rolled by into months, and the mood of the river changed with the advent of heavy rains in the up-country areas. The river floods its banks, and its raging waters uproot trees and gather top soil and bush that float in tangled masses down the river. Occasionally, the bloated, half submerged carcasses of drowned animals pass by, and are carried out to sea by the churning muddy, yellow water.

Gradually, the waters subside and we once more venture out on the river. The body of a really big, dead snake floats by as we paddle upstream. Turning the canoe, we drift alongside it for a short distance, letting the current carry us downstream.

"Do you think it's a python, Chief?" I say, "The bloody thing must be twenty feet long."

The Chief cocks his head to one side, wriggles his nose in distaste, and replies. "The damn thing stinks. Yes, a snake of that size has got to be either a python or a boa. Anyway, they're both constrictors. They crush their prey to death, and swallow it."

Curious, I look more closely. The snake's body, partly submerged and swollen with gases, is predominantly brown with odd lighter shades of brown patterning its entire length.

"Come on, Sparks," urges the Chief, "let's get out of here.

It smells bloody awful." We pull away, leaving the dead snake to float down the river to the sea.

* * *

There is no need for the Chief to yell, "Paddle, Sparks." Like him, I am pouring every ounce of energy into digging deep with my paddle and pulling back hard and fast. Seemingly only yards away is the bow of a Japanese coastal tanker that is bearing down on us. We lie directly in her path as she comes round a tight bend in the river. To our startled eyes, the ship's bow is huge as it towers over us.

The rush of fear lends power to our arms, because we know that, if she hits us, we will be trampled underfoot like matchsticks, and the undertow will push us into her twin propellers. There will be no escape from those menacing steel blades.

The next moments blur in my mind. The image of the ship's bow and the massive bow wave imprint themselves on my senses. The canoe shoots forward, as though trying desperately to escape, and then we are flung up and over by some unknown force. As I go under, I am conscious only of survival, and I try desperately to hold on to the canoe.

Tossed and pulled by powerful currents in this watery world, there is no time for panic or thought. Time stands still, but only seconds have passed when I surface, still clinging to the canoe. The bow wave hurled us clear as it swamped us, and the ship's bow missed us by inches.

I am relieved to hear the Chief, coughing and spluttering as he gasps for air. He, too, has survived and is next to me, clinging onto to the half-submerged canoe as we are swept past the grimy hull. Etched in my memory is a face, a brown face with high cheekbones and the familiar Japanese army cap on his head, looking down at us from the deck of the ship as we scrape by its side. That face and part of the ship's name on its bow, ending in *maru*, is burned into my mind.

As we are drawn towards the stern, I tense my body, fearing the undertow and the suction of the threshing propellers. Thankfully, we are swept clear of them, but it was close, too close. Treading water, the Chief and I hold onto the capsized canoe as we watch the stern of the vessel disappearing upstream. She's is a shallow draft vessel, not unalike the *Luling*. Not expecting to see another ship similar to ours plying these waters, we had crossed the river too close to a blind bend.

"By Christ, Sparks," gasps the Chief. "That was close."

We look at each other, instinctively knowing what we must do. Then, we put the canoe between us and each grip it with one hand as with swim with the other hand, slowly edging our way to the far bank.

That episode leaves us shaken, soaked, but glad to get to the bank in safety. The heat of the sun soon dries out our shirts and shorts as we empty the water out of the canoe. The canoe is undamaged and our spare paddles are still jammed under the seats. Keeping a wary eye on the river, we go on our way.

* * *

Occasionally we use our canoe at the anchorage, but not often, especially after the Chief fell overboard. He was trying to stand up and change positions, but he over balanced. To compound his problems, I ran over him with the canoe when I tried to pull him aboard. The choppy waves made the canoe too unstable, and the Chief had to swim back to the ship.

I think that must be the fastest hundred yards that the Chief has ever swum. I kept close to him, pacing him in the canoe. He swam as if the devil was after him, doing the American crawl all the way, something like Johnny Weismuller in the Tarzan films.

There are sharks in the Gulf. Just the thought of one coming up under you would be enough to spur anyone to swim faster. You can't blame the shark, however, but to them, anything floundering on the surface represents food and is fair game. Wherever there are ships, there is also garbage.

The stewards who work in the galley routinely throw refuse overboard. Naturally this attracts sharks and other fish to the ships. Sometimes sharks follow ships for days in tropical waters. When we were in the Malacca Straits, the crew often put strong fishing lines baited with meat over the stern, and occasionally they even caught a small shark that way.

* * *

Merchant Navy Lapel Badge from World War II

Horace the Ship's Mascot

Ernie's in trouble, and the captain's going crazy. Horace, his pet monkey, has climbed onto the bridge, and was peering at the captain over the bridge bulwark.

"Mr. Mate," the captain yelled, taking a step back, "get this bloody animal off my bridge. I hate the bloody thing. Keep it away from me."

The mate, who has been talking to the bosun on the quarterdeck, was startled, and looked up. Plainly perplexed, Horace climbed on the parapet to get a better look at the Old Man. The captain was going ashore and is dressed in his tropical outfit of white shorts, shirt, knee-length socks and shoes. There was a flash of white as the captain took refuge in the wheelhouse.

Admittedly, the monkey was not small and cuddly, but I liked Horace. I don't know where Ernie got the monkey, but it was probably in the local market in Bangkok. I first saw the two of them walking along the docks about a week ago.

I couldn't help laughing at the unusual pair. Ernie was short and has big ears that stick out. The monkey had long arms and legs, a shaggy ginger-brown coat, and a funny, quizzical face. The monkey obviously adored Ernie.

Ernie held on to the monkey's paw, and the two of them ambled along, completely unconcerned. When I met them at the top of the gangway, Ernie introduced me to Horace. "Look, Sparks, meet Horace," he said, and offered me the animal's paw.

I took it hesitantly, keeping a watchful eye on those long claws and sharp, yellowish teeth. Horace was grinning, with his lips pulled back exposing his gums, not a pretty sight. He grunted and jumped into Ernie's arms. Maybe he thought I'd bite him.

"See," said Ernie, "he's quite friendly. Won't hurt a soul, will you, Horace?" Horace grunted again and picked at Ernie's hair, looking for salt or something.

"Where are you going to keep him, Ernie?" I asked.

"Oh, I'll keep him in the foc'sle head with me and my mates. There's a spare locker in our quarters. Me mates will love 'im. He'll be all right, won't you, Horace?"

Ernie looked at Horace lovingly, then continued, "He's hot, you know. I'd better get him some water. The little fellow's walked a long way. Every now and then I've had to carry 'im." The two of them wandered off down the deck to the foc'sle head.

Horace has settled down with Ernie in his new living quarters and has become the crew's mascot. The main problem was that the captain professed to be scared stiff of him, hence his rapid departure into the wheelhouse.

Attracted by the disturbance, the mate bounded up the bridge ladder. "What's happening, Sparks?" he said. Then spying Horace on the parapet, he doesn't wait for an answer. "Oh, it's you,

Horace, is it? I might have guessed." He made a lunge for the monkey. Screeching wildly, Horace took off and scrambled up the signal masthead. He sat there, jabbering at us, and refused to come down.

Exasperated, the mate shouted from the bridge to Ernie, who was working on the foc'sle head. "Ernie," he yelled, "get this fucking animal off the bridge, and keep him off."

Horace had a mind of his own. He swung in the ship's rigging, defying the crew, until he was finally enticed down by food. Confined to the foc'sle head, he sat on the for'ard deck, forlorn, with a long chain attached to a belt around his waist. His freedom curtailed, he waited for Ernie to finish work.

Soon after Horace's arrival as an unofficial supernumerary to our crew, the ship was ordered to Singapore. Singapore was the naval base and headquarters at that time for the cleanup operations under SEAC (Southeast Asia Command) in the aftermath of the war in the South Pacific.

Early one morning in mid-December, we left Klong Toi and sailed down the Bangkok River for the last time. The *Luling*, no longer limited to convoys after the cessation of hostilities, sailed alone as she headed south, deep into the Sea of Siam, and Bangkok became a memory, a precious memory.

The sea was calm and the twin tracks of the propellers marked our straight, undeviating path through the sea. A faded Red Duster fluttered from a slanting flag pole next to me as I leaned against the ship's rail daydreaming

Hearing voices from the deck below, I glanced down. Long lines of strong cord baited with meat streamed from the ship's stern. Ernie and his shipmates were two decks below me on the open overhang of the short stubby stern. Horace was with him, sitting on the rail with one paw on a cord.

Ernie's voice carried up to me. "That's it, Horace, hold the line." I smiled as I watched the two of them. Ernie was trying to teach Horace how to fish. I wondered what would happen if they really did catch a shark.

"Ernie," I called, "Has Horace caught anything?"

Ernie glanced up and shook his head. "Not yet, but we're hopeful," he said, then looked at Horace. "We will soon, won't we, Horace?" Horace showed his teeth in pleasure and jigged up and down. "See, Sparks, he's gonna catch a big one." He grinned as Horace jumped into his arms to embrace him, his light weight chain clanking against the ship's rail.

Disengaging himself from Horace's clutches, Ernie grinned and looked up at me. "My mates caught two young sharks about three or four feet long, this morning. We killed them and threw them back. You should have seen the water boil as they floated astern. They didn't last long. Those bloody things are cannibals."

The tales of Horace, like those of all ship mascots, are endless because his antics knew no limits. His love for cockroaches was one of his more noteworthy characteristics. The ones aboard our ship were big, two or three inches in length. Horace got loose in the storage hold one day, and went mad chasing and eating them.

Horace remained with us during our voyages around the islands of South East Asia. When we were released from operations with the SEAC (South East Asia Command), Ernie found a home for his monkey with a seaman on the naval base in Singapore. As he handed the monkey over to his friend, the parting is brief. Knowing Ernie and his fondness for Horace, I've no doubt there were tears in his eyes.

* * *

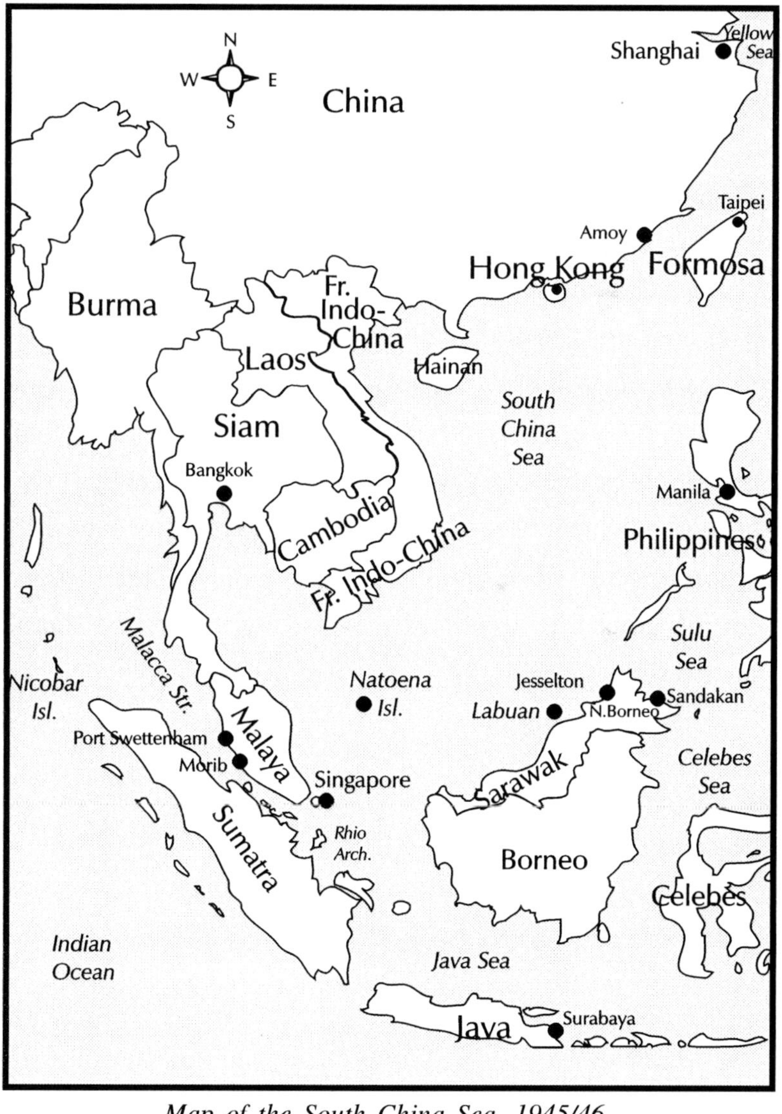

Map of the South China Sea, 1945/46.

Surabaya: The Mini War

The naval authorities don't let the *Luling* stay long in Singapore after our return from Bangkok. We're ordered to Surabaya, Java, with supplies for the naval task force. The assassination of Brigadier Mallaby in Surabaya has led to the occupation of the city, and the deployment of Gurkha troops along a thirty mile perimeter around it. In early January, 1946, the *Luling* is in the bay just outside the harbour.

The heavy crack of gunfire echoes across the bay where we are anchored. Long plumes of smoke wreath the destroyer's for'ard guns. She fires another salvo at the beach, and smoke again erupts from the recoiling barrels of her heavy guns. The blast follows an instant later, and thuds against our ears because she's less than half a mile away.

"Christ, I thought the bloody war was over.," I say to Jim as we stand on the bridge watching the action. " Now look at this lot."

"Remember the refugee ships in Bangkok, Gord?" he says. "They were from ports like this. Java was a Dutch colony before the war. Now we're helping them try and get it back. The destroyer's firing at the Indonesian Liberation Army."

I nod my head. "Yes, I remember them. Mainly Dutch women and children on the ships coming into the anchorage. I wonder where they ended up?"

Jim shrugs his shoulders, "In a refugee camp somewhere, I guess. They were probably sent back to Holland. A lot of them have been here for several generations. There's not much future for them in Java if the Indonesians have any say in the matter."

I'm silent a moment, groping for words. "You know, it's funny how the odd one stands out, isn't it. There was a young woman on board one of the ships. She was beautiful. I couldn't help feeling sorry for her. It doesn't do to stand out in a crowd. It draws attention to you." My voice tails off as I mutter, "The Japanese did such awful things."

We have a grandstand view: the sleek, powerful looking naval vessel passes close to us as it sails through the anchorage. Now shells continue to pound the village. The beach is deserted; there is no sign of life. I scan the shoreline with the binoculars.

Exploding shells hurl thatched roofs into the air and flatten the white-washed adobe huts and houses. There is no return fire. I doubt if there is anybody left alive in the smoking ruins. The destroyer has carried out its support role and retires as quietly and quickly as it came.

* * *

Jim's with me as the naval bus hurtles down the tree-lined road en route to downtown Surabaya. The road from the docks is still damaged from the fighting, and we hang on to our seats as the bus bumps over potholes and detours. There's a bang and clatter as the vehicle swerves violently; the force deposits us on the floor. The driver's cursing. "God-damned fucking pot hole," he swears as he brings the bus back onto the road. We both laugh as we untangle ourselves. The bus seems okay, perhaps somewhat slower, but it's still going.

"Think we'll get there, at least in one piece?" I gasp, recovering my breath.

"Many more holes like that one, and we'll break a bloody axle," Jim replies, then nods towards a little row of Cape Dutch houses. "I bet those old houses could tell a tale." The once picturesque villas are ruined, the gardens overgrown and untended. Large holes gape in the walls, and roofs tilt at crazy angles. They stand in mute testimony to the fierceness of past fighting. "I wonder what happened to the people in them," he adds.

"Well, as you said before, probably in refugee camps, or dead, poor bastards." I stretch my legs out in front of me, and resettle my body in the seat before continuing. "The end of the war didn't solve their problems, did it? Now they're caught in the fighting between the Indonesian Liberation Army and the returning Dutch forces. They can't win."

The city is like an armed camp; the military and checkpoints are everywhere. We're stopped from entering certain streets by military police with the curt explanation: Sorry, sir, there's fighting going on up there.

It's hot and humid in Surabaya; after all, it is January and we're just south of the equator. My khaki shirt is damp with sweat, and so is Jim's. We've spotted a small sidewalk cafe on one of the main streets. It's pleasant sitting here with a cold beer under the shade of a canopy. I feel revitalized. Jim's removed his cap, leaving a red mark round his forehead.

"Too bloody hot and humid," he says, mopping his brow. "God, it's nice to sit down."

I yawn and stretch back, then slouch down in the chair. "Yeah, the trouble with towns is they're too closed in. Any breeze is trapped by the buildings. It's not like this in the countryside."

He laughs as he takes a cigarette out of a pack in his pocket, and lights it. "You'd be in a bloody fine mess out in the countryside here, wouldn't you! The bloody Indonesians would have you for breakfast."

"Hmmm, that's true," I say sleepily, because I must admit I'm tired, too. I shake myself and sit upright. "Well, what d'you think of Surabaya? Bit different to Bangkok isn't it?"

"Gotta give it a chance. We haven't seen the night life here yet, now have we. There's probably a lot of little dives and pubs in the docks and here downtown. The problem is we're not going to be here more than two or three weeks before returning to Singapore. The cargo's half unloaded already."

He sits back, a cigarette in his mouth and admires the passing scenery. "See that young piece passing by. Now I could go for that." His deep set eyes twinkle in amusement as he glances at me, then back at the form of the young Indonesian woman walking away from us.

"Nice bit of stuff, aye, Gord? Do you like that?" he says.

I laugh, because I know the Second Mate's reputation. Maybe, my imagination is not so vivid as Jim's, but she is nice. I like the swing of her hips and the way she carries herself. She's wearing a light, white summer frock with black patterns on it.

It goes with her hair and light brown complexion. "Yes," I say, "I like her hair."

Jim chokes on his beer. "Fucking liar," he says. "I saw you admiring her rear end. That dress is so thin you can see the outline of her panties beneath it." He pauses a moment and continues, "Nice legs, too. You see, Gord, if you met her at one of the local night clubs, you'd see Surabaya in a different light, like Bangkok, now wouldn't you?"

"Yes," I say with a grin, "you're probably right. However, there's not much chance of that if we're only here a short time. Bangkok was different, you must admit. This is just a large straggling port on the shores of a bay with hills in the background. The town is a town, bricks and mortar. Bangkok was quaint, a city of canals and the river running through it to the sea. With a war going on here, it's probably unfair to compare the two cities."

We leave the cafe a little later, and head back to the ship. The naval bus is waiting at its terminal point, and moves off with us aboard at its scheduled time.

Jim is right, our stay is short, about three weeks in all. February sees the *Luling* heading out of the harbour and setting a course for our home base in Singapore.

* * *

Borneo Minesweeping Squadron

Why do I get the impression that the navy doesn't want the *Luling* around? It's late January 1946 when we arrive back from Surabaya and our stay in Singapore is brief. Barely a week later, the Captain announces our next assignment.

"I've just received orders to sail for Borneo," he says at dinner. "We're to take on fuel oil and water, plus general stores. Loading starts tomorrow. We will be joining a minesweeping squadron operating off the Sarawak coast as soon as possible." He looks at the Mate, and says, "I want to see you and the Second to discuss loading schedules after dinner in my cabin."

"Aye, Captain," says the Mate, "we'll be there."

* * *

We join the squadron in a remote area off the coast of Sarawak, dropping anchor in a shallow bay. The foliage ashore is dense, and the long sandy beach is deserted. Except for birds there is no life, and no native boats come to investigate. I look wistfully at the beach, and an urge to explore surges through me.

The Second Mate, a young, stockily built man in his early twenties, laughs when I suggest. "Gord," he says, "You're an incurable romantic, a bloody Robinson Crusoe. There's headhunting Dyaks out there just waiting for you to come ashore. They'll cut your balls off and thread them on a string as a necklace for their womenfolk."

"That's Sarawak, one of the old British colonies," I say with a grin. "As far as I know Dyaks are in the deepest, most inaccessible parts of Borneo. Anyway, the's not a soul around."

"Yeah. You're right," he replies, "but I can't see you convincing the Old Man to run a liberty boat to the shore. We won't be here all that long. The minesweepers will return later this afternoon. They spend the day sweeping up Japanese minefields along the coast, and then come alongside for stores and fuel during the early evening, or morning."

Later that day, the squat grey minesweepers come in. I watch them through the binoculars. The paravanes are stored on the foredeck, either side of the bow. They're strange looking things, like torpedoes, only thicker in the body. They have strange shaped wings, or fins, designed to keep them at a fixed depth when they are towed beneath the surface.

I've seen the paravanes in operation before, off the Scheldt Estuary in the North Sea. They are effective against moored mines floating just beneath the surface in shallow coastal waters. The towing cables have serrated edges and are attached to the lowest part of the bow. As the paravanes are pulled through the water, they fan out on either side of the minesweeper. The wire cuts the cable holding the mine and releases it. The mine then floats to the surface where it is destroyed by gunfire.

There are four ships in the squadron. For a while our crew is busy as one by one the minesweepers come alongside to take on fuel or stores. Because we're only a couple of degrees above the equator, there are no long summer evenings or short winter days here. Sunset varies only fifteen minutes or so, on either side of six in the evening. Tonight, the sun has long gone from the evening sky when the last ship departs.

The lights of the anchored minesweepers glow in the dark across the bay. There's no moon tonight, but the sky is lit by stars. It's after dinner and a solitary light bulb on the bulkhead casts a soft yellow light over the boat deck. Only the distant hum of the ship's generator breaks the peace of the night. The deck chair creaks as I lean forward, my hand reaching for the half empty glass on the wicker table. The beer, golden yellow and topped with froth, sways as I pick up the glass. It tastes good, cold, but not icy cold, because it's been out of the refrigerator for ten or fifteen minutes.

Jim's face is in the shadows as we talk. He looks towards the minesweepers, and nods as I ask about their operations. "Yes," he says, "I've been talking to the deck officers on board them. I was wondering about that, too; I mean, about the type of mines they're clearing."

He pauses, then continues. "They're mainly moored mines close in shore, but there are magnetic and acoustic mines, too. The majority are Japanese, but some of the mines are American and British. They were probably laid in counter-mining operations by aircraft or submarines."

Mine laying aircraft! That has a familiar ring. I remember the Scheldt Estuary and Holland in late 1944. No sooner had the minesweepers cleared the channels by day than German aircraft laid more mines at night. The thought of ratchet mines makes me shudder. The allies never found an answer to that one. Those pressure operated mines caused devastating losses during the last year of the war. I only hope there are none in these waters. My last ship was blown in half by one of those mines in the Scheldt Estuary, Southern Holland

The minesweepers operate from the bay. As the minefields are cleared, we move on along the coast. It's less than a week since we arrived and the busy little sweepers have cleared the first sector of coastline of mines. The squadron is off to a new operational base. The *Luling* sails out of the bay and sets a northeasterly course along the coast of Sarawak, heading for Labuan, a small island off the coast of northeast Borneo.

There's little traffic on the radio, just weather reports. The coast stations have no messages for us. I wander out of the radio cabin onto the bridge. Ahead of us, the minesweepers are in formation. Their combined wake trails intermingle and spread out uniformly across the calm surface of the ocean. The coastline on our starboard side shimmers in the heat haze. Like an unwanted child, the *Luling* tags along behind.

I smile to myself, my arms resting on the bridge weather shield as I watch them, their white ensigns fluttering from their sterns. I prefer the casual life of the merchant fleets to that of the Royal Navy.

A voice startles me out of my daydreams. It's Jim; he's just come out of the wheelhouse and joins me on the wing of the bridge. "A penny for your thoughts, Gord," he says with a laugh. "You solving the world's problems?"

"Christ, some hope. Afraid my thoughts aren't worth much," I say, as I half turn my head. "Just thinking how different we are to those stuck-up naval guys. Life's more free and easy on merchant ships. I like it that way."

Jim nods, "Yeah, me, too. I suppose you've got to have discipline in a fighting force, but the Royal Navy takes it too far. Probably it's got something to do with the old class system. Maybe that'll change now the war's over."

My thoughts go back over the past as we talk, and memories come back. I reply bitterly. "They'll never change, at least the older generation. The class system is instilled into the English. It's a part of them, like the caste system in India."

"In a way, I feel that I've lost my teen years." I shake my head in frustration. "Since 1942, I've been in war zones. How can you meet a nice girl and take her out like any normal teenager? I'll be twenty-one soon. There are no social clubs available for merchant seamen, not in war zones. The only women we meet are the ones in the dives and dance halls. What the hell do they expect us to do? We're not bloody monks."

Jim laughs at the thought. "That's true. Well, personally, I don't give a shit. Society can think what they like. I'm certainly no bloody monk. It's hard being thrown into this lot when you're young." He shakes his head. "How anyone could opt for a life of celibacy beats me. Maybe they haven't got balls."

"Humph," I say bitterly. "I was brought up by nuns and

monks, the Brothers of the Order of the Sacred Heart. I hate the lot of them. They're a bunch of hypocrites. Brother Jarlath, my homeroom teacher, liked little boys. I liked and trusted him until he touched me. After that I steered clear of him. He molested several boys in the school before being found out. All they did was transfer him to another school. Can you believe that? He should've been jailed, the bastard."

My outburst is a rare because normally I keep these thoughts to myself. The events of the past few months, especially in Bangkok, have been preying on my mind. The intensity of my feelings moves Jim. I feel his hand on my shoulder. "You're dead right, Gord," he says. "We're living in a sick world."

* * *

The draft from the whirling fan in the chartroom catches Jim's hair and blows it over his forehead as he looks down at the map, and he automatically smooths it back. "There," he says, his finger tracing the pencilled course on the chart. "That's Labuan. It's an island about six miles off the coast. We'll be there later this afternoon, about 1600 hours. The coastline we passed a short while back was Brunei. It's a tiny sultanate sandwiched between Sarawak, and north east Borneo."

The minesweepers are already in port as the pilot boards us outside the small port of Victoria. First impressions are often deceptive, but as I watch from the boat deck, I'm not impressed by what I see. The main downtown area itself lies some way from the docks, and merges with them in an untidy slum of housing.

The harsh discordant ring of the engine room telegraph on the bridge reaches my ears. The *Luling* moves ahead slowly through the harbour mouth. The wash of her bow wave ripples across the still water to splash against the rocky breakwater guarding the entrance. Now that we are closer, I can see the warehouses lining the docks in greater detail. They are in varying states of disrepair, some badly damaged by past fighting. The minesweepers lie at anchor off shore, a few hundred yards from the docks.

There is an increase in activity on the bridge, and instinctively I glance up. The bridge telegraph has rung, and the throb of the engines has changed note. The familiar call, "Stand by, Mr. Mate," rings out from the bridge to the Mate and his crew on the foc'sle head. We anchor near the minesweepers.

* * *

There's a gap in the thornbush hedge, and behind it I see jagged metal glinting in the rays of the sun. My curiosity aroused, I peer through the gap. The scattered remains of a Japanese aircraft lie just beyond, half-hidden in the scrub and grass. The round red insignia of the rising sun on the body and wings of the aircraft are plainly visible.

Aircraft fascinate me. I joined the Air Defence Cadet Corps when I was eleven back in 1936. Now as I pick my way through the gap in the hedge, I step over broken strands of barbed wire, all that is left of a once high security fence.

The scattered wreckage of Japanese aircraft litters the derelict landing field. In the distance, I see the scorched, burnt-out hangars and a control tower.

It brings back memories of 1939/40 when I served with the air cadets on an airfield in southern England. I remember Taffy, a leading aircraftsman who was an armourer attached to the Northweald Fighter Bomber Squadron. I was fourteen, and was his winger, helping him to load the hurricane fighter aircraft with machine gun bullets and test the guns on the firing range. The government took us cadets off the airfields in the spring of 1940 when the German planes started bombing them.

As I inspect the crumpled remains, I wonder what tales this plane could tell. It's a single seat fighter, probably a Zero, its fuselage badly burnt and twisted. The cockpit is blackened, and its once transparent, sliding cover is bent and hanging off.

The aircraft is not in an earthen bunker, like they used at the airfield in Northweald. *Perhaps there was a surprise bombing raid, but there are no bomb craters.* I sit down on the wing with the ghosts of the past. How long I sit there, I don't know.

My thoughts go back to the *Luling*. Jim was busy this morning when the liberty boat left, so I decided to explore the countryside alone.

Absently, I pick up a piece of the cockpit canopy, turning it around in my hand. *Pity,* I think to myself, *Jim would have liked to have seen this*. I drop the discoloured metal on the ground, and glance at my wristwatch. It's nearly four in the afternoon. The liberty boat leaves the quay at five.

Reluctantly, I get to my feet. The late afternoon breeze stirs the overgrown grass on the airfield. This is a peaceful place, quiet like a grave. With one last look at the aircraft, I retrace my steps through the strands of barbed wire and undergrowth. My shoes scuff the dust on the dirt road as I wander back towards the docks and ship.

* * *

Labuan is disappearing into the heat haze astern of us as the *Luling* ploughs eastwards through calm seas. We're en route to Jesselton further along the north coast of Borneo. It's a little less than a hundred miles away. At eight knots the *Luling* should make it in ten or eleven hours. The minesweepers have left us and are disappearing over the horizon to the east.

As I come off watch, the Captain's on the starboard wing of the bridge. He's young for a Captain, in his late twenties or early thirties. His eyes are glued to binoculars as he scans the harbour entrance to the small port of Jesselton on our starboard beam. He turns as he hears me approach. "Any news of the pilot boat, Sparks?" His voice is low and soft.

"Yes, Captain," I say, handing him the latest message. "I've just been in contact with the port authorities; he's on his way."

He glances at it and strokes his chin reflectively. "Good," he says, then looking up, smiles. His eyes sparkle; they're large and intense, eyes that hold you. "We should see the boat any minute now." He grunts with satisfaction as he focuses the binoculars, and picks out the pilot boat just clearing the harbour.

* * *

We're anchored near the minesweepers, two or three hundred feet off the main jetty. The pilot's gone now, and the *Luling* sways on the anchor with the tide. It's hot and humid. My shirt's sticking to my back. Irritated by sweat bees, I wipe my eyes as I look towards the town of Jesselton. From where I stand on the boat deck, I can see the red tiled roofs of the buildings. They are partly masked by the docks and warehouses.

The tiled roofs along with some roofs made of corrugated iron sheets, appear hot. Maybe it's just the colour, but the thatched roofs of nearby native housing seem cooler..

The anchor has been down barely fifteen minutes, and the crew are stripping off their clothes. Naked, they plunge from the tank deck into the cool, inviting depths of the surrounding sea. It's deep and the water's crystal clear, except for the intense reflected blue of the sky.

The Aldis lamp on the nearest minesweeper flashes frantically. I hurry to the bridge. Jim is with me as I answer its call. The first word—*Sharks*—jolts me as I translate the message aloud for the Second Mate: *Advise no swimming in these waters.*

Jim wastes no time; he yells to the bosun. "Get those fucking men out of the water, Bosun. There's sharks!"

The effect is electrifying. Just the word, *shark*, is sufficient to clear the sea of swimmers. Men swim frantically for the Jacob's ladders hanging down the sides of the ship, then haul themselves up the ladders to stand, dripping wet, in little groups on the deck.

Evidence of the presence of sharks becomes abundantly clear a few minutes later. As I lean on the ship's rail, my eyes searching the water below me, I see a movement in the depths. The noise of splashing objects on the surface has attracted the curiosity of a hammerhead shark. The two eyes on each end of the wide hammer-shaped head look grotesque, but the graceful movement of its huge body through the water as it curves up to the surface is beautiful and at the same time menacing. I watch in awe as it turns in a shallow loop and sinks back again into the depths from where it came.

The men clustered on the tank deck below watch in silence. Then, as it disappears, a crewman says to his mate, "Fucking hell, Jack. That's no tiddler. One bite from that bloody thing could kill you."

His mate agrees. "Aye, lad," he says. "Yon shark could've swallowed us up in one bloody gulp. Fuck that for a lark."

* * *

We left Singapore over a month ago in early February. Since then the tireless little minesweepers have been clearing mines along six hundred and fifty miles of coastline. They've swept from the western limit of Sarawak to the Balabac Strait that separates Borneo from the Philippines. It's March 25, 1946, and the *Luling* is heading for the strait, a hundred miles north east of us.

We are en route to Sandakan on the shores of the Sulu Sea, our last base before we return to Singapore. The coast, a thin distant green fringed line with mountain ranges in the background, lies on our starboard side. Jesselton is but a memory and is disappearing fast behind us to the west.

The minesweepers are already clearing the strait, the most northerly part of northeast Borneo. Soon they'll head southeast to Sandakan where we'll meet them.

Night has fallen as we reach the Balabac Straits, and I'm still within radio range of both Jesselton and Sandakan coast stations. I've worked them periodically throughout the day. Now as I'm going off watch for the night, I call up Sandakan radio.

The Morse key chatters, using the jargon of the sea, the "Q" code. *QTH—position.* I answer: *Balabac Straits. QRU—have you anything for me?*

The reply is swift and efficient: *QTC GN—nothing, good night*. I make a note in the radio log and sign off.

Leaning back in the chair, I stretch and smile to myself. How different life is aboard ship compared to the war years. *In just a short period of time, my job has changed drastically.* Now I can transmit the ship's radio traffic round the world through coast stations or direct, using our HF transmitters. Radio beacons on the coasts are being restored, and with them, the use of our direction finding equipment. The old order has changed and I like it. The radio cabin has taken on a new significance for me, and I'm proud of my station. Life is good.

I wander out on the bridge. It's Jim's watch, and I chat awhile to him. The ship's movement as it ploughs steadily through the night sets up its own breeze, and, although warm, it feels pleasant on my face.

There is a distant flashing light on our port for'ard quarter. The Second Mate nods towards it. "That's Balabac Island light beacon on Cape Mellville," he says. "There's lots of bloody islands around here. Thank Christ, they've restored the beacons. We can at least take cross bearings on them."

The remark amuses me and I laugh, because I had similar thoughts just a short while ago. "You know, Jim," I say, "in the short time since the war ended, our jobs have changed drastically. No convoys and station keeping. Light and radio beacons are commonplace. We can transmit traffic freely. It's great. Just an hour ago I was talking to Sandakan, and earlier today to Jesselton. It's a different world."

"Thank Christ for that," he replies.

* * *

The pilot's aboard and he's taking us into harbour. It's mid morning and the pilot boat follows in our wake through the narrow entrance between towering cliffs. Set against the backdrop of mountains and luxuriant vegetation, it is one of the most beautiful natural harbours that I have seen.

Thick bush covers the steep lower slopes of the rocky cliffs as the harbour opens out into a large deep-water creek. Hills and ravines surround the small town. The business area is close to the docks and, from the ship, appears to be one long, main road lined with open-fronted shops. The roofs of the buildings are mainly thatch, in contrast to the corrugated iron or tile of more modern structures.

Surrounded by the razed remains of former residential buildings, the ruins of a burnt-out church stand like a lonely sentinel in the background, just above the main part of the town.

We anchor in the harbour. The minesweepers are not here. The pilot told the Captain that they went off early today to sweep along the coast. We did not see them as we sailed down through the Sulu Sea from the straits, but then we stayed further out to sea. They will be back this evening.

As the sun dips towards the west, the grey minesweepers file through the harbour mouth . The *Luling* is busy, and there's a minesweeper on either side of her now as she lies at anchor. The sound of the pumps throbs into the early evening as bunkering continues, until, like little sated pigs suckling a sow, they've had enough and slip away to sleep at anchor in the bay.

* * *

"You from ship?" The voice is hesitant, unsure, the English poor, but then my Chinese is nonexistent. A few words make all the difference.

I grin and nod, then say, "Yes, from ship. We arrived yesterday." The sentence is too lòng and difficult. As I talk his features relax. Then he, too, smiles.

The harbour lies several hundred feet below the ridge on which we stand. My uniform shirt and shorts are dirty from scrambling through ravines and clambering up cliffs. When I reached the ridge, I accidentally came upon a house in small clearing. As I walked out of the thick bush onto the grassy lawn, I was surprised to see the house with its neat and tidy garden, and I must have startled the small family group.

The man is taller than the average Chinese. He talks to his wife, a short, middle aged woman, and then turns to me. "This wife." Pointing to the two young girls he adds, "Two daughters." They are all smiling now as he continues, "You like drink?"

How can I refuse? The sweat is showing through my shirt and on my face. The heat during the middle of the day is too hot for comfort, and climbing through the bush has sapped my energy. They laugh as I accept the offer eagerly.

A white linen cloth is laid on the ground and a tray with cups of tea is carefully placed on it. We all sit cross-legged on the lawn under the shade of a flowering tree, and converse in pantomime, gestures, and occasional words. The young teenage girls, both dressed in European style clothing, are delighted at this new game.

When one of the young girls asks my age, I reply, "Twenty," while I hold up my hands and open and close my fingers and thumbs twice.

The results are hilarious as the two girls try to tell me their ages. One is sixteen and the other eighteen.

Regretfully I rise to my feet, point at my chest and the distant harbour. "Must go," I say. "Is there an easy path down?" I pantomime an easy slope with my hands, and then and shake my head as I indicate a steep one.

The father's face lights up. "Path easy! I show." He takes me along a narrow winding path through the bush to where a dusty road descends from the ridge. I thank him, and set off down the road to the harbour. I can hear the high pitched voices chattering excitedly as the father and his daughters turn back through the bush.

* * *

I wish our stay was longer, but we'll be going in a few days. Our mission is completed. The coast from the Balabac Strait in the north to Darvel Bay in the south, bordering Dutch East Borneo, has been swept clear of mines. The minesweepers head for Singapore, while far behind the fast-moving naval vessels the slow moving *Luling* ploughs north through the Sulu Sea at a sedate speed of eight knots. Once clear of the Balabac Straits, we set a southwesterly course across the South China Sea for Singapore.

* * *

A Voice from the Past

The *Luling*, a seventeen-hundred-ton tanker, ploughs steadily through the South China Seas, roughly three hundred miles east-northeast of Singapore. Far astern, the Natoena Islands seem to float upside down in the heat haze on the distant horizon. The ship's bell strikes the change of watch, and the spell is broken. Reluctantly, I turn away and make my way to the radio cabin.

The small merchant ship resembles a boot, because she has a short stubby foredeck, and all her accommodation is aft. At first glance, she seems an unlikely candidate for the Royal Navy but she has been carefully selected. Her shallow draft enables her to operate with ease in the rivers and irrawaddy of Burma and Malaysia, where larger vessels cannot go. That's why she was chosen as a supply ship to a naval assault force gathering in Trincomalee, Ceylon, for the invasion of the Malayan mainland.

The beachhead landings in August 1945 are now just a memory. The *Luling*, along with the invasion force, was held off the Nicobar Islands for over two weeks while the Japanese surrender was negotiated. Now it's early April 1946, and we're involved in cleanup operations as part of the takeover phase.

Today, the heat and humidity in the radio room are oppressive. I've just finished transmitting a long shopping list for the ship chandlers, as well as routine reports to the port authority and naval headquarters using the working frequency (475 kHz). Gingerly, I move the earpieces of my headset slightly to one side. My ears are tingling because the earphones have been clamped on my head for over an hour.

The Singapore coast station has signed off and returned to the calling frequency (500 kHz), and I'm about to follow suit, when I hear a ship calling me. *Damn*, I think, mentally cursing the bloody ship. *It's probably someone unable to contact Singapore and requesting me to forward their transmission, but it could be something urgent*. Curious, I reach for the Morse key and make contact.

As I listen to the Morse code streaming through the headphones, I can hardly believe my ears. *Hi, Gordon, this is Jim, Jim Hankley. We sailed together*. He was the Chief and I was his Second Radio Officer on the *Empire Path* in December 1944. The realization brings me fully alert in my chair, and I listen intently, copying the message.

Jim is now the Chief Sparks on a ship that is in the anchorage outside Singapore harbour, along with John, who had been the Third Sparks on the *Empire Path*. They've invited me to have dinner with them on board their ship when the *Luling* gets into Singapore, and will send a boat to collect me.

There are still so many questions that I want to ask, but the answers will have to wait until I see them. The voice from the past has caught up with me and, with it, the memories.

The last time that I saw Jim was in Ostend, Belgium, on Christmas Eve, 1944. Along with the rest of the survivors, I was lined up on the dock when they took the wounded, including the Chief, in the ambulances to the hospital. He'd been injured by flying debris when the *Empire Path* was blown in two by a mine off the Scheldt Estuary, Holland. Mike, a young engineer and a close friend of John, the Third Sparks and me, lies dead among the wreckage of the *Empire Path* on the bottom of the North Sea, barely fifteen or twenty miles off Ostend on the Belgian coast.

Shaking my head to clear away these thoughts, I concentrate on the weather reports now being broadcast by Singapore. There is no change in the forecast. Tomorrow will be another calm and balmy day in the South China Sea.

It's early evening as I come out of the radio cabin onto the bridge, and I see the Second Mate's dim form on the starboard wing. "Jim," I call softly, unable to conceal the excitement in my voice. "I've been in contact with my old Chief Sparks; he's on a ship in Singapore."

He turns, and comes towards me, as I tell him about the message from my old Chief, and the dinner invitation. "That was your last ship before you joined us, wasn't it?" he says, his voice quickening with interest. "Were many killed?"

"Considering that the ship was blown in two, there were surprisingly few deaths." I pause briefly as I remember that day in December 1944. "The engine room crew on duty were killed outright," I say slowly. "A ratchet mine exploded under the hull amidships. They didn't stand a chance."

Jim nods, "That's true. There's not much between you and

the sea down there, maybe half an inch of steel plate."

"Yes," I say, my voice choking. "There were five men there, including Mike, a young junior engineer. Mike used to go out with the third Sparks and me whenever we were in port. We met his sister just before we sailed at a dance back in 1944. She was in the Women's Auxiliary Forces (WAAF) stationed in an army antiaircraft unit not far from the docks."

I clear my throat, and then continue. "Strange thing was that I saw her again walking with an elderly naval officer about a week later in Fenchurch Street. She was crying. I had arrived home the night before, and gone up to London to Siemens head office. I should have gone across the street to try and comfort her, but I just could not. Events were still fresh in my mind, and I knew that I would burst out crying."

Pausing to control myself at the vivid memory, I look away and then add, 'We were all about the same age, about nineteen."

The memory is still vivid in my mind, and I talk in halting sentences. "The boilers blew up, too. A lot of the crew were injured, including the Chief Sparks, but the rest of us were lucky, I guess."

Jim glances down and shifts his weight to the other foot; his mind is far away. "I know what it's like. So many young fellas died," he says softly.

Thinking of Mike, I bite my lower lip. "The explosion was terrible, like a giant fist through the centre of the ship. I was in my cabin at the time, and got out onto the bridge. The engine room housing was peeled open like a tin can, and the ship ripped in half. All the lifeboats were destroyed except for one, and I managed to get off in it. A minesweeper picked up about six or seven men from her bow as the ship was going down."

Closing my eyes momentarily, I can see the ship's stern as we rowed away. The propeller is clear of the water. The ship's back is broken and both ends are rearing up. Two men are jumping from the ship's stern. I tear my mind away from the memory and the scene fades back into my subconscious.

"Two men who were left aboard jumped into the freezing water and were picked up by an escort vessel." Feeling calmer now, I speak slowly. "Counting the thirty aboard our lifeboat, and the ones taken off the bow, as well as the two in the water, that's roughly forty survivors that I know of. It happened around four o'clock on Christmas Eve, and it was still daylight."

Jim's face is thoughtful, "Yeah, it's surprising not more were killed. Who picked you up?"

"An RAF Air-Sea Rescue launch, a converted motor torpedo boat," I say, smiling as I remember the packed launch. "The MTB was stationed out in the North Sea to pick up downed fliers from bombing raids on Germany. We were the largest boatload that it had ever carried."

* * *

The following afternoon we reach the anchorage. Singapore lies off our starboard bow as the *Luling* swings with the tide on our anchor chain. I've given Jim Hankley our time of arrival, and told him that the *Luling* is a small coastal tanker.

In anticipation, I make my way to the bridge where I can scan the ships at anchor with the binoculars. The flashing of an Aldis signalling lamp attracts my attention. It's beamed on us from a Liberty ship. *It's Jim. He's spotted us. They're sending their ship's liberty boat to fetch me.*

When the boat comes alongside the *Luling*, I swing my leg over the deck railing and scramble down the Jacob's ladder. Both Jim Hankley and John are in the boat to greet me. Jim's not changed, the same big burly frame and bluff features. He's in his late thirties or early forties.

The Chief is beaming at me as he stands on the lifeboat seat holding the end of the swaying rope ladder. "Christ, Gord," he says, as he helps me into the boat, "it's good to see you again."

I laugh as I reply, "How the hell did you know I was on the *Luling*?"

"That's easy. I'd asked the company about you, and they told me the name of your ship. When I heard you working the coast station, I recognized your call sign."

John grins a welcome. "Hi, Gord," he says. "What are you doing on that old tub?"

"Same as you, by the looks of it. We're attached to a naval assault force operating out of Singapore," I add with a laugh. "The only trouble is that we don't stay long in port before they give us another assignment. We've just returned from North East Borneo where we acted as a supply ship for a minesweeping squadron."

My thoughts wander as I look at John. He is just like I remembered. He's about my age, maybe a year older. His wavy brown hair is swept back from his forehead, and his grey eyes sparkle when he smiles. He reminds me of Mike, full of life and fun, only Mike's dead.

We're heading directly for a Liberty ship riding at anchor,

and John's pointing at it. "That's our ship, Gord, the *Harold. B. Watts*. Sounds impressive, doesn't it. A seven-thousand-tonner. The accommodation's good, too, but you'll soon see for yourself. We'll be there in a minute."

A crewman stands in the bow with a boat hook and steadies the boat as we draw alongside the bottom platform of the gangway. Jim leads the way as John and I follow him up to his cabin on the boat deck under the bridge. Deck chairs and a table are already set up on the wooden deck, and an awning provides shade from the late afternoon sun as it sets in the west.

"Ah, that's better," Jim says, taking a sip of the ice cold beer. "The first today. Come on, Gord, there's plenty of beer. What have you been up to since we last saw you on the *Empire Path*?"

I pour the beer gently into my glass. It's an American brand and has an unfamiliar taste. It's not as strong as the local beer, but I like it.

"It's a long story, Jim," I say. "You know some of it already, but first I'm curious to know what happened to you after you went into hospital in Ostend. About the only thing that I know of your travels is that you and John have probably just come from the States. This American beer is a dead give away."

"Yeah, we stocked up in San Francisco. That's about five months ago. Stocks must be getting low now." He laughs and adds, "Do you like the beer?"

I nod, "Yes, the taste is different. It's a bit weaker than I'm used to, but it's cold and relaxing." I stretch and lie back in the deck chair, settling my body comfortably in the canvas seat. "Were you long in hospital, Jim?" I ask. "The Mate told us that you'd got hit by flying debris and broke your collar bone. Is that right?"

"Yes," Jim says. "It all happened so quickly. One moment I was sitting in the chair in the radio room, and the next I was slammed against the bulkhead. A heavy metal box—the one that we put the decode manuals in—hit me in the face and shoulder. The bloody thing nearly knocked me out."

He shakes his head, "The radio room was a shambles. They patched my face and collar bone up in the hospital, then sent me home about two weeks later." He pauses a moment, then continues in a lighter vein. "It's a pity you didn't sail again with John and me. We were sent over to the States a year ago, and spent more than three months in New York."

"Yeah," John interrupts, "we were there so long that I got a part time job as an elevator operator and bellhop in a hotel."

Jim laughs, and says, "We had a great time. The company paid for the hotel and our board, and then sent us by train across the continent to San Francisco take over the ship. The bloody ship was not ready for commissioning, so we stayed ashore, all expenses paid, for another three months. We were in New York for VE day and San Francisco for VJ day. How about that, now?"

I shake my head as I think of the irony of it. I too, could have gone with them. I look at them as they sit on the deck chairs against the backdrop of the lights of Singapore. Night falls quickly in the tropics and it's dark already.

"Well," I say, "I had the opportunity to come with you, but I wanted to have my own ship." I look up at the starlit sky. "There's someone up there who has something else in mind for me. It's thc way it is, I guess. Anyway, where the hell have you been since you left San Francisco? Trading in the Pacific?"

Jim glances at John, "It must be five months now, John, since we left the States."

"About that," John says, then adds, "We left for Australia in early October."

Jim thinks a moment before answering. "Yes," he says. "We took cargo down to Sydney and Perth, then did a few odd trips to New Zealand before we came over here to Singapore. We were damned lucky to hear you working Singapore, Gord. It's quite a coincidence."

Seeing Jim and John again is like meeting up with family. We were together for only a month on the *Empire Path* before it was sunk, yet it feels as if a lifetime has passed between us. They are old friends. I nod my head and feel a lump in my throat as I reply, "Yes, Jim, it's strange that our paths should meet again like this, and in Singapore of all places. It means a lot to me, just to see the two of you." I lapse into silence, unable to find the words to express my feelings, but I know they understand. We've gone through so much together.

Jim's face softens and he empties his glass. "It's as if it were meant to be." He looks up as he puts his glass down on the wicker table. "We are working our way back to Europe with general cargo. We're off early the day after tomorrow, via India and the Suez Canal."

We talk on awhile, chatting over old times. The sun set some time ago in the west over Sumatra; and it's time to go down to the wardroom for dinner. I can't help contrasting their ship with ours. It's like a close knit family, while ours tends to be disruptive, broken into groups. Perhaps that's because we're a small coaster and have been constantly thrown together for over a year now.

As I'm leaving, Jim shakes my hand and says, "Gordon, I want to thank you for contacting Siemens, and telling them that I was in hospital in Ostend. They were on the point of sending a telegram to tell my family that I was missing. It saved them from a lot of pain and hurt to know that I was safe."

I study Jim's face as I reply, "You were in hospital in Ostend, and I felt that someone had to report and that it should be me. I had arrived home about midnight the previous day, but we don't have a telephone. That's why I caught the early train to Woolwich and saw Siemens personally." I continue talking as we turn and walk along the deck to the gangway. "The company had just been notified that our ship had been sunk, and they were astonished to see me walk in. It felt strange."

He says softly, "Yes, I should imagine it was." Jim laughs briefly, then says, "I was lucky, I suppose. That metal box is a bloody heavy thing. It's fortunate that my shoulder took the brunt of it, and the cuts on my face and head were superficial."

We stand at the top of the gangway, while the boat waits below. I look down at the deck, make up my mind, and look at Jim. "Maybe I made the wrong decision. The company wrote to say that you'd asked them to keep us together on our next ship, but I wanted my own station."

I pause a moment, choosing my words. "I don't really understand why I felt that way, but coming back on the Royal Navy LST, there was no one to speak for John and me. We could understand the lack of accommodation in the wardroom, but being denied the privilege of eating our meals there, especially after being told by the Petty Officer in charge of catering that we could do so, was humiliating."

Looking up at Jim, I say, "I'm sorry, Jim. I know if you'd been there, it would have been different. But I was bitter and determined that never again would I be humiliated like that. I needed the rank of Chief Radio Officer to restore my own esteem. That's why I asked for my own ship."

I feel his hand on my shoulder, and there's a smile on his face. "I heard from John how you both were treated aboard the LST. Believe me, Gord, I would have felt the same way. You know now what it's like being Chief and you're doing fine."

I leave them there, the two of them, Jim and John, as I walk down the gangway, and enter the boat. I had asked them to come for dinner with me on the *Luling* the following night, but time is against us. They are sailing in a little over twenty-four hours.

The Liberty ship *Harold B. Watts* fades into the night and, as the boat heads across the anchorage, the familiar shape of the *Luling* looms out of the darkness. The old ship is everything I have. Despite her faults, my heart warms to her. I'm home again.

The *Luling* lies in the anchorage, swinging round its heavy anchor chain, at the mercy of the tides. The Old Man's gone ashore in his whites with the Mate to a naval conference in Singapore. Late in the afternoon they return and disappear into the chart room on the bridge.

That evening, barely two weeks since my twenty-first birthday, at dinner in the wardroom, the Captain, almost casually, says:

"The Borneo minesweeping operation was our last mission for the navy. The *Luling* has been returned to her normal peacetime status as a coastal trading vessel."

There is dead silence for a moment as realization sets in. Then, he is besieged with questions. Having got our attention, he continues. "The *Luling* is being taken over by one of the big China coast shipping companies for the balance of our two-year contract. As you already know, this tanker is part of the American lend-lease package to China under the Marshal Plan, and is scheduled to be handed over to the Chinese Nationalist government under Chiang Kai Shek in Shanghai."

He pauses, and sips his coffee. "We will be based in Hong Kong and will be trading up and down the coast of China. At least that's the general plan."

* * *

Adolph S. Ochs

I joined the *Adolph S. Ochs* on January 24, 1947, in London and sailed under Captain Tyson. The war was over. There were no more convoys, just the natural elements to fight. The Atlantic had plenty of those, with heavy seas, gales and driving rain squalls. I started using *mothersil* on this voyage. This seasick pill had been developed from research into sea sickness among troops on beachhead landings during the war, and had not been available for purchase until then. That pill was my saviour; I was not sea sick once during the entire voyage.

Our destination was Saint. John, New Brunswick, Canada. Little did I think that Canada would eventually be my home, but that would be some time in the future. The sea voyage across the Atlantic took two weeks, and when we approached our destination, thick fog blanketed the ocean.

* * *

The mournful sound of the ship's fog horn echoed as the Liberty ship *Adolph S Ochs* forged ahead at reduced speed, pushing blindly through moderate fog-bound seas. As I came off radio watch, Captain Tyson was pacing the bridge, sextant in hand, hoping for a break in the fog. The captain, although a very capable man, had been badly shell-shocked during the war, and never seemed to stay still for a moment. He had a nervous jerk to all his movements.

The time was twelve noon, but there was no sun for the captain to take the ship's midday position. He turned towards me. "Aye, Sparks," he said, his voice betraying his Scottish background. "Bloody fog. I canna see the bloody sun to take a bearing for all this weather. Have ye the latest weather report for the Bay of Fundy?"

"Fog, sir. The whole coast of Maine and the Bay of Fundy, right up to Saint John is fog bound. Looks like we'll have to use the direction finding equipment and take cross bearings on the coastal radio beacons."

The captain had already ordered a watch to be kept for'ard on the ship's bow for other vessels in the shipping lane and fishing craft. Returning to the radio cabin, I started to take radio bearings on the Sable Island radio beacon about 300 miles out in the Atlantic. Captain Tyson had been on a naval vessel during the war, and put great reliance in his electronic navigational equipment. That was unusual, because most Merchant Navy captains and mates at that time were slow to adjust to the changing world.

Our normal speed would be nine or ten knots an hour, about average for vessels on the high seas in 1947. We ploughed on through the fog, our siren periodically sounding its long mournful note, the ship moving slower now that we were approaching Cape Sable.

After we passed Cape Sable we came in range of other radio beacons, and I continued to take radio bearings every few hours. Now I was able to get cross fixes using other coastal radio beacons on the Nova Scotia coast in Canada and Maine in the USA, enabling us to plot our location accurately. I was employed by Siemens, and our direction finder was a Siemens rotating loop.

Still in dense fog, we entered the Bay of Fundy moving slowly up through its waters while I checked our positions on the direction finder equipment using radio beacons on the New Brunswick and Nova Scotia coast. We were sailing through fog for about four or five days before we arrived just off Saint John. The captain was kept busy with his charts, drawing the changes in course and distances run all this time, while I operated the direction finder equipment. He and I worked closely together for several days as we coordinated the ship's positions.

We were especially busy when we were nearing the Port of Saint John. As I gave him the last bearing on the Saint John lightship radio beacon, the captain turned to me and said, "Well, Sparks, let's go out on the bridge and see if our reckoning is right. The lightship should be coming into sight any moment now." Sure enough, we had no sooner gone onto the bridge than the lightship loomed up out of the fog, its siren greeting us.

I was surprised at the accuracy of our readings on the DF and very pleased. It also renewed my confidence in the equipment and the captain's confidence in my ability to take accurate bearings.

The pilot boat came alongside, and, with the pilot safely aboard, we entered the harbour and docked alongside the wharf. The tidal change from high to low tide in Saint John amazed me. It must have been at least thirty feet. We docked at high tide and appeared to tower above the dock, but at low tide one had to look up to see the dock edge. The main deck was well below the dockside and the steel cables and hawsers had to be slackened off periodically and later tightened, depending upon the state of the tide.

Thinking back, I don't remember too much about the city of Saint John, except that the snow was very deep and the temperature was well below freezing. I have a vague memory of a main street and some large department stores. I bought nylon stockings for Joan and her sisters, as well as other little gifts for family in one of the stores before returning to the ship.

Unloading and loading ships was a fairly lengthy business at that time, and on general cargo ships such as ours it would take anywhere from two to three weeks or even more. The ship was fully laden about the end of February, and we left Saint John to head back across the Atlantic to Liverpool.

In 1941/1942 the Kaiser shipyards in New York alone were turning out Liberty ships at a rate of one a day to overcome the great shortage of merchant ships, caused by heavy losses during the Second World War. The early Liberty ships were designed to meet an immediate need, and were not designed to last more than a year or two.

Most Merchant seamen looked at Liberty ships with some degree of scepticism. These ships soon developed the reputation of breaking their backs and sinking in the severe weather. Welding ship plates together formed a very strong bond at the joint, but softened the metal on either side of the weld. In heavy weather, the flexing of the metal hull could crack the hull along one side of the weld, and the ship would sink. Although more advanced and robust Liberty ships were build later in the war, their reputation had been badly tarnished.

Personally, I liked the Liberty ships for the roominess of their accommodation and comfort. My cabin on the *Adolph S. Ochs* was the most comfortable and pleasant cabin that I had during my time at sea. The furnishings were excellent, and windows (not port holes) looked out onto the upper deck of the amidships accommodation.

Inside the cabin, there was a bed instead of a bunk, a writing bureau, a comfortable chair, and a settee, as well as ample cupboard and drawer space. The cabin was well lit and pleasingly decorated. When I joined the ship, she had been in service for four years. The *Adolph S. Ochs* had been built by Bethleham-Fairfield Shipbuilders, and launched in 1943. Despite the welded hull, the ship was in good shape and did not appear to experience any problems.

* * *

Blurred and indistinct through the rain and mist, the coast of Ireland took shape on our port bow, and then disappeared. As we entered the Irish Sea, we set a new course for our home port, Liverpool. Picking up the pilot off the mouth of the Mersey River, we slowly moved into the river mouth and docked alongside in the Birkenhead docks.

The usual routine customs and immigration procedures were completed, and the crew were paid off on March 15 in the wardroom. The captain called me aside, and asked me if I would like to sign on again for the ship's next voyage. This was peacetime, so the Ministry of War Transport no longer dictated which ship you joined. My employer, Siemens Bros. Ltd., had contractual services agreements with the shipping company, so it was up to me whether I join the ship or not.

At that time, British Merchant Navy radio officers were employed by one of three companies—Marconi, International Marine Radio, and Siemens. These companies supplied both the equipment and the radio officers under contract to various shipping companies, although a few shipping companies employed their own company radio officers. I was quite happy to sign on again the ship for another voyage, so I signed the ship's articles, packed my bags, and then hurried to the train station, heading for London, my wife Joan, and home.

Joan and I were living with her mother, who had a nice little home in Chingford, a suburb of London, until we could get a place of our own. Joan worked as a secretary in a law firm in Walthamstow. Living in her mother's home was very convenient for Joan, especially when I was away at sea.

My leave was all too short, as leaves invariably are, and three weeks later I caught the train to Liverpool Street Railway Station in Central London en route to the Liverpool docks. Early in April 1947, the *Adolph S. Ochs* sailed for Argentina. Ireland and the Fastnet Lighthouse were soon behind us, as we sailed steadily south through the Atlantic, and the cold rainy weather of the North Atlantic gradually grew warmer with the passage of each day.

Headed through calm seas and under blue cloudless skies, we sailed south past the Madeira Islands, then the Canary Islands, and Cape Verde Island, not stopping even for bunkering or supplies. We crossed the equator and into the South Atlantic with only minor changes of course. Six hundred miles out of Buenos Aires, I contacted the coast station and gave them our position and also advised the ship's agents of our estimated time of arrival. Radio communications were surprisingly good as the contact on 500 kHz was on the extreme edge of our transmitter's range.

We entered the Rio de la Plata, passing Montevideo on our starboard bow, and headed up this wide estuary of the River Plate, to Buenos Aires. The pilot boarded in the estuary and we slowly moved into the city's large harbour to a waiting berth.

No sooner was the ship tied securely to the dockside bollards and the gangway lowered, than hordes of officials and dockyard workers sprang aboard. The hold covers were taken off and the protective planking and metal girders removed by the ship's steam-operated winches. The bustle and activity of unloading cargo began in earnest.

Buenos Aires itself was a lovely city. One of the deck officers and I went into the downtown area, not far from where the ship was docked. A pub called the *First and Last* was located near the dock entrance: that was probably as far as some merchant crews got on shore leave. We passed the pub and strolled out into the city along wide streets, passing large department stores and both modern and older buildings. There were many squares of varying sizes, with at least one monument in each of them. I remember one very large, attractively landscaped area crisscrossed with wide roads: I believe there was a large car park built underneath it.

The *Mission for Seamen* in the downtown area was always a good stopping place, and it was one where you could meet local people. We stayed about two weeks in Buenos Aires. Then, with our cargo unloaded, we left Buenos Aires early one morning and proceeded about 100 miles up the River Plate, or more correctly, the Parana River.

As we travelled up the river at reduced speed to prevent the ship's wash from damaging the river banks, we passed through flat open plains known as pampas, dotted here and there with the little villages. Late in the afternoon we reached Rosario, the second largest city in Argentina. The port had large dock facilities for ocean-going vessels, and the pilot gently eased us into a berth alongside the dock.

Perhaps more so than Buenos Aires, I will always remember Rosario because of something that happened there. Looking back, I cannot remember whether I had been to a movie at a local theatre that night or just wandered about the town, exploring the local shops and night life. In any case, it must have been about eleven or twelve at night when I returned to the ship. The docks were dimly lit with a few strategically placed security lights illuminating the way between packing cases and other material unloaded from the ships.

I had just turned the corner of a large warehouse, when suddenly I was surrounded by a pack of stray dogs of all shapes and sizes. The experience was frightening. These packs of stray dogs came from the Pampas surrounding the city, and often roamed the dock areas looking for food. They kept circling me, while I tried to edge towards my ship. When the dogs began to attack,

I shouted and kicked out at them. One or two dogs retreated, but the other dogs dashed in and bit my legs.

Fortunately for me, the barking and commotion plus my shouts aroused the dockyard watchmen who came running to my rescue and drove the dogs off. When I got back to the ship, blood was dripping down my legs from numerous bites. Luckily, none of the bites was too deep. With the help of the guard on the gangway, I put antiseptic ointment and bandages on the deeper cuts, and went to bed.

When the captain heard about the bites the next morning, he contacted the shipping agent, and I was taken to the local *Louis Pasteur Clinic* for rabies treatment. From the long lineup of local people at the clinic, dog bites seemed to be quite common in Rosario. Rabies was very rife in the Argentine, and, if you could not identify the dog or dogs that bit you, then you were automatically given a course of antirabies injections.

Rabies was very serious, and treatment, which was traumatic, had to start as quickly as possible. The first day, four shots of rabies vaccine were injected into the stomach area. This was followed by a course of injections, tapering down to one shot per day after the first week. The shots of vaccine were injected into the buttocks. All the time that we were in Rosario, I had to attend the clinic daily to have these injections.

When the ship left Rosario two weeks later, the task of giving injections was turned over to Captain Tyson. He wasn't very good at giving injections, and I distinctly remember that he stuck the needle into my buttocks while he filled the syringe. Fortunately he handed over the task to his second in command. Regularly every morning, I went to the captain's cabin to get my injections, until, thankfully, the month-long course of injections was completed. Having gone through such a painful and traumatic ordeal is sufficient reason for anyone to remember a town.

During this episode, we sailed downstream to Buenos Aires and anchored in the harbour. A liberty boat was organized from our ship to get people ashore for shore leave and business purposes, and get them back to the ship again. Quite by chance, I was on our liberty boat when I met an old school friend, John Newton, in the harbour. It had been years since I'd last seen him.

Our liberty boat was just about to pull away from the landing dock, when the liberty boat from another ship drew up alongside. There wasn't too much time to talk, except to briefly exchange news because of the liberty boat time schedules. John was attached to another ship in the harbour, but I'm not sure whether he was a cadet or apprentice deck officer. Meeting him was a pleasant surprise, especially in such unusual circumstances. Shortly afterwards our ship sailed for the UK.

This was my last voyage in the Merchant Navy. According to my Discharge Book, we arrived in Birkenhead, Liverpool, on June 27, 1947. I signed off the ship's articles and returned to London and home.

* * *

Index

Acknowledgments

All the events in this book are factual, and are recorded as I remember them. Throughout the book, actual names are used for the ships; however, some personal names have been changed.

I wish to acknowledge The Society of Authors as the Literary Representative of the Estate of John Masefield for permission to quote from his poem, "Cargoes".

The cover photo (View from aircraft of a convoy of merchant ships assembling in the North Atlantic) is from the National Archives of Canada, PA-115005.

The map of the South China Sea has been based on artwork from One Mile Up Library.

Family photographs and other illustrations are from my personal collection of photographs and memorabilia.

Thanks to my fellow authors in the BC Marine Authors and the Vancouver branch of the Canadian Authors Association and other writing groups for their support and encouragement.

A special thanks to Barbara, my wife and partner, not only for encouraging me to write, but also for her hard work and assistance in making things happen.

* * *

About the Author

Gordon Mumford was sixteen when he began his marine radio training, and seventeen when he went to sea as a Merchant Marine Radio Officer in World War II. In 1949, he went to East Africa as a field services engineer, working on a one-thousand mile VHF radio-repeater route. This project was a worldwide prototype for long distance communications.

He immigrated to Canada In 1958, and studied at the University of Western Ontario (where he met Barbara) and the University of Toronto. After returning to Kenya in 1961, they spent twenty years on international projects in Kenya, Nigeria, and South Africa. Gordon was a technical instructor, while Barbara worked as a librarian or a teacher.

The Mumfords returned to Canada in 1980 with their two sons, settling in Vancouver. Gordon worked for Communications Canada as the federal government emergency planning liaison officer, and did two short teaching contracts in Saudi Arabia and Indonesia.

He began writing books after he retired in 1995. *The Black Pit...and Beyond* (2000) and *The Sampan Girl* (2001) were published by General Store Publishing House. *White Man's Drum* (2004) and *Drums of Rebellion* (2005), both about Africa, and *Dangerous Waters* (2008) were published by Zebra Publishing House, while *In Fate's Footsteps* is nearing completion.

Websites: http://www.gordonmumford.com
publishing@gordonmumford.com